Tokyo Phantasmagoria

An Analysis of Politics and Commodity Capitalism in Modern Japan Through the Eyes of Walter Benjamin

Kenny Loui

Dissertation.com
Boca Raton, Florida

Tokyo Phantasmagoria: An Analysis of Politics and Commodity Capitalism in Modern Japan Through the Eyes of Walter Benjamin

Copyright © 2008 Kenny Loui

Dissertation.com
Boca Raton, FL
USA • 2008

ISBN: 1-59942-676-5
13-ISBN: 978-1-59942-676-1

Tokyo Phantasmagoria

"Kenny Loui's *Tokyo Phantasmagoria: An Analysis of Politics and Commodity Capitalism in Modern Japan Through the Eyes of Walter Benjamin* is a careful and deeply thought application of the ideas of German-Jewish philosopher Walter Benjamin to the context of postwar Japan. Loui looks at the way that history is understood and remembered in Japan through the lens of Benjamin's musings about the way that notions of time and history are themselves products of capitalistic manipulation, what Marx calls commodity fetishism. In the case of Japan, Kenny shows how the idea of time itself is employed as a way to create a certain sense of inevitability and destiny in terms of current power arrangements, in the process erasing or subsuming the experiences of the war and other practices that are more problematical. In his application of Benjamin to the Japanese context, Loui demonstrates a tremendous gift for translating a western thinker into a non-western context. This work mixes a vivid description of a 'Tokyo Phantasmagoria' that evokes Benjamin's own writings on Paris in the 19th century with a serious critique of postwar ideology and social practice. Loui's turn to Benjamin helps to allegorize and render legible the functioning of government, relationality and art in today's Japan."

—James R. Martel, Associate Professor and Chair,
Department of Political Science, San Francisco State University

"Tokyo Phantasmagoria: An Analysis of Politics and Commodity Capitalism in Modern Japan Through the Eyes of Walter Benjamin is a timely analysis of a Japan that is struggling with national identity in a time of economic stagnation, globalization and changing demographics. Bringing the analytical structure of Walter Benjamin to bear in an interpretation of post-WWII Japan, Mr. Loui illuminates the role of government influence in the historical interpretation of events, specifically of sexual slavery in Japan's era of expansion and war in the early 1900s, and deftly connects it with the general treatment of women in 21st-century Japanese society. He highlights the complacency of the public and the dangers it faces from disinterest, falling under the mollifying influence of consumer society, and the modern world's harsh quantification of human qualities. His reframing of Benjamin's solutions of art and allegory (in a Japanese context) as a way out of the confines of blind faith in the consumption ideal offers an insightful solution to combating the trappings of affluence, while retaining a unique cultural identity. Japan is struggling with the transition from fast growing, savings-driven economy to a stable, mature economy, and Kenny expertly points out how Japan can either act as a cautionary tale or a guiding light to other countries following the rapid-industrialization trajectory."

—Brent Burgess, Centre for Environmental Policy and Governance,
London School of Economics and Political Science

ACKNOWLEDGEMENTS

This thesis would not have been written if I had not taken Professor James Martel's post-modern political theory seminar in the spring of 2006. Thus, I would first like to express my thanks to Professor Martel for introducing me to the writings of Walter Benjamin and for actually persuading me to take on this project. I must also commend Professor Martel on his exceptionally fast turnaround times while reviewing my drafts! And of course, the advice and suggestions he gave me throughout the writing process was invaluable (especially the suggestion on including an analysis on Takashi Murakami's art in the final chapter). I am also grateful for Professor Martel's support and guidance since my first semester in the political science graduate program at SFSU.

Secondly, I owe a debt of gratitude to Professor Gerard Heather—my advisor and mentor—without whom I would not be where I am today. He was one of the few who saw my potential when I was an undergraduate student, encouraging and supporting me in all my academic and professional pursuits. It was an honor not only to have been one of Professor Heather's students in three separate classes but also to have been one of his graduate teaching assistants during the fall 2005 semester.

The idea of linking Benjamin's writings to contemporary Japan would never have come to me if I hadn't gone to Japan in the first place. Therefore, I would like to thank Mr. My Yarabinec, Ms. Kati Bell, and Ms. Marisa Thigpen, who introduced me to the option of studying abroad in Japan, and ultimately encouraged me to do so via SFSU's Bilateral Exchange Program. My year abroad was a worthwhile experience that not only provided the impetus for this project, but helped me to redefine my goals in life.

During my academic year at Mejiro University in Tokyo, I enrolled in Professor Misao Makino's "Japan Phenomenon" course and Professor Ignacy-Marek Kaminski's "Japan in Asia" course. Several of the topics I studied in these two classes subsequently became prominent subjects of inquiry in this thesis. Hence, I would like to say *arigatō gozaimasu* to both Professor Makino and Professor

Kaminski for piquing my interest in and adding to my knowledge of many of the prominent issues in Japanese politics. I also want to thank Professor Masahiro Kubodera, with whom I took a seminar on the topic of Japanese love stories, for our engaging one-on-one discussions and his insightful reflections on Japanese youth and the state of traditional values in contemporary Japan.

I would also like to say *merci* to my "big brother" and *sempai*, Mr. Brent Burgess, with whom I had the pleasure of studying at Mejiro University. He is one of the most intelligent individuals I have ever met. I am thankful for his encouragement and kind words when I was working on this thesis and for believing that I could actually finish it in time for a May 2008 graduation. At the time I was writing the thesis, Mr. Burgess was busy with his studies at "the second best school in the world" (which I'm sure moved up the ranks to number one the moment he set foot on campus), so I did not want to burden him with any extra work but I did ponder the idea of recruiting Mr. Burgess to be an unofficial third member of my thesis review committee. As he is a "resident expert" on the two cities I compare in this thesis, having lived in both Paris and Tokyo, I'm certain that he would have offered me brilliant insights and invaluable feedback.

Of course, I have to thank my parents who have always given me their love, guidance, and support. Specifically, I thank my father for showing me the value of diligence and perseverance. He has worked so hard to support me since Day One, to give me the academic and professional opportunities that he was never given, and for that, I am truly grateful. I hope that my accomplishments are something that he can be proud of. As for my mother, I think she is the kindest woman in the world. (Personally, I think my father is a very lucky man. If I am ever as fortunate as my father to meet a woman with my mother's heart—her capacity for good, her selflessness and self-sacrificing nature—I would marry her in a heartbeat.) She has my utmost gratitude for drilling into my very soul the importance of altruism and human compassion. I am truly blessed to have the both of them in my life. Even though we've had our fair share of disagreements and trivial quarrels over the years, I want my parents to know that I love them very much.

Lastly, I would like to acknowledge Obi-Wan Kenobi of *Star Wars* fame, whose words provided the guiding principle for and one of the overarching messages in this thesis, as he told Luke Skywalker that "truths" are essentially subjective and open to interpretation, that "many of the truths we cling to depend greatly on our own point of view."

TABLE OF CONTENTS

INTRODUCTION

Revisiting the Cave

Plato's allegory of the cave, as presented in Book VII of *The Republic*, tells the story of humans imprisoned in a dark cave since childhood. All they see are the shadows of puppets presented on the wall in front of them. These shadows are, to the prisoners, reality as they know it. Plato states that "such persons would certainly believe there were no realities except those shadows of handmade things" (Rouse 1983, 313). Imagine if one of the prisoners were released and then exits the cave, exposing himself to the outside world beyond the shadows. The truth, unbelievable as it is, is revealed to him and he comes to the realization that what he originally thought was reality is nothing more than an illusion. That is Walter Benjamin's goal: To lead people out of the cave, to dispel the "phantasmagoria" of mythic reality that people have been exposed to since birth.

Walter Benjamin, an early 20th-century German-Jewish philosopher and literary critic, whose written works examine topics including, but not limited to, history, art, society and politics, was a profound thinker of his time. His writings, most notably "The Task of the Translator," "The Work of Art in the Age of Mechanical Reproduction" and his incomplete study on the 19th-century Paris arcades, complied after his death and published as *The Arcades Project*, present thoughts and commentary for Benjamin's generation, which was experiencing the rise of both commodity culture and fascism. Benjamin put forward ideas—as well as signs and portents—relevant to his time, but what do the writings and social commentaries of a man who lived during the early 1900s, and whose life project focused primarily on the 1800s, have to do with the present day? In other words, are Benjamin's works still relevant today? In short, the answer is yes. This discourse on the writings of Walter Benjamin will illustrate that humanity is still trapped in the cave—the cave of commodity-capitalism.

Applying Benjamin's works to 21st-century Japan, connecting significant social and political issues facing Japan in the present and in the years to come, will show how his writings are universal, being both relevant and applicable to various epochs and civilizations. Japan, Tokyo in particular, has been chosen for this study because of its rapid post-war technological and economic advancements and the rarely addressed detrimental consequences of those advances. Though Japan is said to have entered the modern age in the time between the two world wars (Harootunian 2000, x), it essentially had to start from scratch and "re-modernize" after World War II. Thus, Japan's post-modernization is a relatively recent phenomenon. As such, the symptoms of commodity-capitalism that consumed 19th-century Paris, the primary focus of Benjamin's *Arcades Project*, can be seen in its infantile and adolescent stages in 21st-century Tokyo.

Reviewing every written work by Benjamin is beyond the scope of this discourse and thus it will be limited to a few selections of Benjamin's writings. Specifically, the writings that will be evaluated focus on the "dream image" of reality as *perceived* by the average person living in a commodity-capitalist society. The emphasis on the acquisition of wealth, to the point of adopting materialism as a value system, feeds into this dream image, warping people's perception of the world in which they live. As a result, people not only see what they want to see but are susceptible to seeing what *others*—particularly government and corporate entities who harness the power of the media and other tools of propaganda—want them to see. As a result, the populace finds itself trapped in what Benjamin refers to as the "phantasmagoria," an illusory reality of a nearly-flawless society defined by technological advances and economic prosperity. But this phantasmagoria of modern life shrouds the "grotesqueness" of underlying truths such as the failure of advances in science and philosophy to eradicate poverty and other social inequities—instead the phantasmagoria hides these social ailments—and the transformation of humans into the very commodities that they value. For that reason, even though this thesis is not about prostitution per se, the topic is examined several times throughout the essay because of the link prostitution provides between humanity and commodity: every wage earner, selling her or his services, is in essence a prostitute in that the prostitute is simultaneously the seller and the commodity. As Benjamin states in his exposé of 1935 of the dialectical (i.e., two-sided) nature of the prostitute, "Ambiguity is the appearance of dialectic in images, the law of dialectics at a standstill. Such an image is afforded by the commodity per se: as fetish. Such an image is presented by the arcades, which are house no less than street. Such an image is the prostitute—seller and sold in one" (Benjamin 1999, 10). In the end, Benjamin wrote appealing to his own generation, but as this discourse on modern-day Japan and the "Tokyo phantasmagoria" will reveal, Benjamin appeals to our generation as well.

Fragments

Much of Benjamin's writings reference fragments in one way or the other. In a similar manner, each chapter of this thesis can be considered a separate and distinct fragment that can either stand alone or be combined into a mosaic—or as Benjamin would say, a *montage*—depicting political and social life in contemporary Japan. The first chapter itself deals with the combination of seemingly distinct "shards" of Benjamin's written works into a coherent and illuminating discourse.

Chapter 1, "Translation and Interpretation," focus on three of Benjamin's essays: "The Task of the Translator," "The Work of Art in the Age of Mechanical Reproduction," and "Theses on the Philosophy of History." At first glance, these stand-alone works may seem unrelated in content, but they shall be taken as fragments of a larger whole, that once put together, reveal the importance of definitions and the power of interpretation in a political society. These essays will be analyzed alongside the issue of comfort women, the thousands of women who the Japanese Imperial Army forced into sex slavery during the Second World War. This issue has been chosen not only due to its relevance in Japanese politics today, but because it provides an actual real-world exemplar of what Benjamin discusses in his writings of how a government can create and proliferate to the public a certain view of reality, which may not necessarily be a lie, but may not necessarily be the entire truth either.

Chapter 2, "Tokyo Phantasmagoria," draws upon quotes and commentary from *The Arcades Project* to elucidate the illusory nature of the modern city (specifically Tokyo) in masking the problems within capitalist societies, such as poverty. The chapter will also discuss how commodity-capitalism in Japan, with its emphasis on materialism as a value system, has humanized the commodity while at the same time, commodified humanity. Primary topics of focus will be the Japanese sex industry's commodification of the human body and the Japanese education system's commodification of the human mind. In this chapter, Benjamin's project of fragmented text on the Paris arcades will be, to use his terminology, exploded out of history and into our own time as the commodity culture of 19th-century Paris is compared to that of 21st-century Tokyo.

The final chapter, entitled "Of Art and Allegory," builds upon the discoures presented in the previous chapters. The first chapter discusses definitions and interpretation, while the second chapter smashes the false reality of modern life in Japan, illuminating the phenomena that lie hidden within the mist of phantasmagoria. Chapter 3 focuses on the application of art as a means for people to engage in politics and the use of allegory as a device to render false what was once perceived as truth and in so doing, provide a weapon for the citizenry to combat against government or corporate propaganda, and against the often unjust and oppressive 'status quo.' The unquestioned "wisdom" of Confucian values will be used to illustrate the use of allegory to shed light on a phenomenon's dialectical nature, revealing its hidden aspects.[1] That Confucianism is usually taken as dogma in Japanese society is the very rationale for why it will be the focus of allegory for the purpose of showing how deeply held beliefs can be shattered by revealing its hidden dark, or as Benjamin would say, grotesque, side. Finally, the "Superflat" style of contemporary Japanese artist Takashi Murakami shall be explored, showing how Murakami utilizes allegorical imagery in his art, in a manner not unlike how Charles Baudelaire employed allegory in his poetry, as a means of criticizing— deliberately or not—Japanese commodity culture and its consequences.

The thesis will culminate with a discussion on the Messiah, a figure of *hope* in Benjamin's writings, and political action by the citizenry as a means of dispelling the phantasmagoria. The purpose of this political-philosophical discourse is two-fold: To explore Benjamin in the context of contemporary Japanese society and, via the Japanese case, to get readers to think deeply about and perhaps even question their own perceptions of reality in relation to the social problems and political challenges most, if not all, modern industrialized nations face today. As stated above, Benjamin wrote for his generation, but his thoughts and words are still applicable to our generation today and perhaps to future generations in the decades and centuries to come.

The Methodology of Montage

The analytical method Benjamin uses is that of montage, with the intent purpose not of describing, but of *showing*. "Method of this project: literary montage. I needn't *say* anything. Merely show. I shall purloin no valuables, appropriate no ingenious formulations. But the rags, the refuse—these I will not inventory but allow, in the only way possible, to come into their own: by making use of them" (Benjamin 1999, 460; N2,1). What can be seen in Benjamin's montage of quotations and commentary? That is up to us—for each individual reader—to decide, for there is no one correct interpretation, no one truth. Just as he discusses interpretation in his writings, Benjamin leaves us with a literary montage for us to interpret with our own eyes. In the end, a commentary on reality is an interpretation of that reality: "Bear in mind that commentary on

---

[1] The word 'dialectic' (or 'dialectics') has several definitions. One definition is: "A method of argument or exposition that systematically weighs contradictory facts or ideas with a view to the resolution of their real or apparent contradictions" (*American Heritage Dictionary* 2004). Drawing from the above definition, for the purposes of this essay, I will define 'dialectic' as contradictory properties inherent in an object or phenomenon. To illustrate my point, take the following example from Benjamin: "Dialectic of flânerie: on one side, the man who feels himself viewed by all and sundry as a true suspect and, on the other side, the man who is utterly undiscoverable, the hidden man" (Benjamin 1999, 420; M2,8). Is the man visible or invisible? He is one or the other, and simultaneously both—that is how he is "dialectical." These dialectical properties, i.e., contradictions, can exist simultaneously in something, yet at times, one aspect of the object or phenomenon is promulgated as "truth," while the other aspect is rarely acknowledged or completely ignored. Allegory can be employed as a means of bringing into the fore this hidden aspect.

a reality (for it is a question here of commentary, of interpretation in detail) calls for a method completely different from that required by commentary on a text" (460; N2,2).

Reading Benjamin is ultimately an interpretive and thought-provoking process. Benjamin urges us to suspend the truth—what we perceive of as truth—and to cultivate our imaginations to read not only between the lines, but beyond them. *The Arcades Project* is not a book per se, but a collection of quotations and Benjamin's commentaries and anecdotes. These literary fragments, of course, can be put together much like a puzzle to discover the messages that Benjamin is trying to convey. As such, this work is written as a constructed puzzle (or a montage); actually, more like two different yet interrelated puzzles combined, with fragments from Benjamin's writings fitted together with fragments of relevant issues and events from contemporary Japanese political and social life. In the concluding remarks to *The Dialectics of Seeing*, Susan Buck-Morss's study on Benjamin's *Arcades Project*, the author encourages readers of her text to engage in an "interpretive project" of their own on Benjamin's colossal yet incomplete work (1991, 340). This thesis is one such project, one that will advance the argument that commodity-capitalism exhibits similar effects regardless of time or place and that the past and the present are not so different, essentially being one and the same—what Benjamin would refer to as the "eternal return of the same" (Benjamin 1999, 71; B4,1)—and it is for these reasons that the writings of a man who grew up in Germany during the early 20th century but wrote primarily about life in 19th-century France can be applied to events and phenomena occurring in 21st-century Japan.

CHAPTER 1

Translation and Interpretation

"Truth, for all its multiplicity, is not two-faced," wrote Charles Baudelaire in *Oeuvres* (Benjamin 1999, 315; J48,3). Related to the topic of truth and its multiplicity is the following observation Walter Benjamin makes on fashion:

> On the publicity war between the fashion house and the fashion columnists: "The fashion writer's task is made easier by the fact that our wishes coincide. Yet it is made more difficult by the fact that no newspaper or magazine may regard as new what another has already published. From this dilemma, we and the fashion writer are saved only by the photographers and designers, who manage through pose and lighting to bring out different aspects of a single piece of clothing ..." Helen Grund, *Vom Wesen der Mode*, pp. 21-22. (72; B5,1)

As "different aspects" can be discovered by looking at one piece of clothing by changing the positioning or varying the lighting, different aspects can be revealed by observing one piece of historical "fact," not by altering lighting, but by altering one's position, i.e., one's point of view. Much of what we perceive of as reality is highly dependent on our own perceptions and points of view. And in turn, what we believe is often times *learned*, both in formal and informal settings, be it in the classroom or through our own personal experiences and interactions with the world around us. History is one particular form of knowledge that cannot be acquired via our own personal experiences—it must be taught to us. The key questions then are: What is being taught, or *not* being taught, and from whose perspective?

A major political issue Japan has been facing since the end of World War II is its perception of wartime history, which has caused a rift not only between Japan and other Asian nations, but a rift among the Japanese themselves. The Japanese government's stance on comfort women—the government's inability to take *full* responsibility for, and at times, even the flat-out denial of, the military's institutionalized sex slavery prior to and during the Second World War—is a point of controversy that has caused on-again-off-again tensions in Japan's diplomatic relations with China and South Korea. Even Japanese historians debate among themselves the validity of the historical "facts" behind the comfort women issue. Overall, these domestic and international dilemmas are a result of conflict between opposing *perceptions* of history. Using a selection of Walter Benjamin's short essays and a few excerpts from *The Arcades Project*, I shall examine how history is not necessarily a series of objective facts strung together, but a narrative that contains truth as well as elements of "mythic reality." As Benjamin himself states, it is his goal to dissolve—to awaken others from—this mythic reality, and this awakening occurs when one is conscious of past events that have been covered up (or erased) by the historical narrative presented for her or his consumption: "While in Aragon there remains an impressionistic element, namely the 'mythology' [...] here it is a question of the dissolution of 'mythology' into the space of history. That, of course, can happen only through the awakening of a not-yet-conscious knowledge of what has been" (1999, 458; N1,9).

Benjamin presents a quotation on the past's "purity": "To read into the future is difficult, but to see *purely* into the past is more difficult still" (1999, 470; N7,5). Benjamin then supplements with a brief commentary: "The 'purity' of the gaze is not just difficult but impossible to attain" (470; N7,5). Ultimately, this impossibility at grasping the purity of the past is due to the subjectivity people, be it government officials or historians, apply to their interpretation of historical events. Benjamin's "The Work of Art in the Age of Mechanical Reproduction," "The Task of the Translator," and "Theses on the Philosophy of History," as we shall see, contain topics that are interrelated and deal with interpretation and perception of reality and their importance in the realm of politics. First, Benjamin's discourse on film technology in "The Work of Art in the Age of Mechanical Reproduction" is presented alongside a government's ability, in the role of "cameraman," to propagate a certain perspective of reality; in the Japanese case, the national government's ability to present a specific standpoint of wartime history to the general public. Second, the distinction between "mode of intention" and "object of intention," as discussed in "The Task of the Translator," will be applied to the case of comfort women to show how a government can utilize language and conceptions to its advantage to make the public believe its "official" point of view. In relation to the topics presented in "The Work of Art in the Age of Mechanical Reproduction" carrying on the analogy of government as cameraman, we can consider modes of intention—i.e., words—to be the government's roll of film. Third, a brief discussion on "Theses on the Philosophy of History" brings everything together, showing how fantasy and reality are ultimately combined to form the "truth," and how the citizenry are strung along like puppets to believe in this one particular perspective on events—namely, the government's perspective.

This discourse on reality and one's definition and perception of reality will ultimately serve as a prelude to the next chapter, which focuses on the phantasmagoria, a much deeper illusion than that of the historical narrative. To elaborate, there are many people who are consciously aware of the Japanese government's alteration of historical truth—hence the ongoing debate within government and academia about the national government's official view of history and what "really happened"—but there are fewer people that will debate—let alone are aware of—the phantasmagoria. Like historical narrative, the phantasmagoria is essentially a narrative of its own, but one that does not rest in the past but permeates throughout time immemorial. In short, like "official" history, the phantasmagoria is an illusion derived from a certain point of view, but its effects are on a much larger scale, for practically anyone and everyone in a commodity-capitalist society—from a corporate manager or high-ranking government official to a supermarket cashier or high school student—is influenced by it. In other words, in contrast to the case of Japanese wartime history, the illusory nature of the phantasmagoria—of life in commodity-capitalism— is rarely, if ever, disputed or even acknowledged as such. Ultimately, the average person lives and breathes the phantasmagoria each and every day without ever realizing there is a hidden reality behind the one that she or he experiences.

Definitions and Historical Narrative: The Sovereign's Camera

"The Work of Art in the Age of Mechanical Reproduction" presents Benjamin's thoughts on the advancement of tools to reproduce art and the repercussions of such an advancement. In Section I of the essay, Benjamin gives an overview of the evolution of art reproduction, from founding and stamping to the motion picture. Section II puts forward the concept of authenticity as it relates to art. "Even the most perfect reproduction of a work of art is lacking in one element: its presence in time and space, its unique existence at the place where it happens to be" (Benjamin 1988b, 220). The history of a work of art—its creation, where it has been, where it is—is the indicator of its authenticity, of its *uniqueness*. "The presence of the original is the prerequisite to the concept of authenticity. [...] The whole sphere of authenticity is outside technical—and of course, not only technical—reproducibility" (220). Or is it? As methods of reproduction progressed, mass production of art became possible and subsequently widespread. As a result, since a work of art may be in multiple places at once, for practically everyone to see, authenticity ceases to be of importance or of significance to the artwork. Benjamin remarks, "By making many reproductions it substitutes a plurality of copies for a unique existence" (221). Mentioning the authenticity of photographs, Benjamin asserts, "From a photographic negative, for example, one can make any number of prints; to ask for the 'authentic' print makes no sense" (224). And once authenticity is deemed virtually irrelevant, art no longer has a basis in 'ritual,' meaning its location and original usage, but, as Benjamin articulates, a basis in *politics* (224).

Among the modes of reproduction in the modern era, it is the motion picture that has the most political importance. This significance lies in film's ability to construct reality. Benjamin explicates how reality is pieced together by a camera through the comparison of a stage actor and a screen actor. When actors act on stage, they do so in real time before a live audience; their actions, as the scene progresses, are logically sequential. Screen actors, on the other hand, act in front of the camera and scenes need not be filmed in chronological order. The fragments of footage are then spliced together in the appropriate order during postproduction and then presented to the audience as a completed sequential movie. The order in which the scenes were filmed is irrelevant to the audience, who sees the events of the film in the temporal sequence in which they were meant to be presented. Concerning film acting, the camera is ultimately in control, not the actor. "The audience's identification with the actor is really an identification with the camera. Consequently the audience takes the position of the camera" (Benjamin 1988b, 228).

In political terms, one can make the analogy that a sovereign (be it a king, emperor, president, prime minister, congress, parliament, or other governing authority) is the cameraman and the citizenry is the audience, and in the same way that "the audience takes the position of the camera" (Benjamin 1988b, 228), the citizenry takes the position—the perspective—of the government. Like the cameraman, the sovereign controls what his subjects see; the subjects see what the sovereign wants them to see. Just as the audience views a movie through the filmmaker's eyes, the citizenry views reality from the sovereign's eyes. Like pieces of footage, fragmented events may be reorganized and reassembled in a way that serves the sovereign's purposes. Benjamin asserts, "What science has 'determined,' remembrance can modify" (Benjamin 1999, 471; N8,1). Like science, historical events themselves are "determined," but as one—in our particular case, the sovereign—"remembers" these events (even if he has never experienced them first hand), he has the ability to modify them via his historical narrative. For example, one perception of the Japanese Imperial military during the decades of Japanese expansionism is that of conquerors, while another is that of the military as liberators. Victims of Japanese imperialism have argued the former perspective, whereas certain groups within the Japanese government have supported the latter claim, that the Japanese were "liberating" the rest of Asia from the threat of U.S. and European colonialism.[2] Like today, the early 1980s saw revisionism in Japanese history textbooks. In 1982, the language used in textbooks to describe the Imperial Army's activities in China during the 1930s used words like "advances" instead of "aggression" to China's dismay, who saw the textbook revisions as Japan's attempt "obscure the nature of their wartime behavior" to the point of "denial that they had committed atrocities" (Hayes 2005, 229). Ultimately, reality is whatever the sovereign reconstructs it as, and whatever he is able to make the public believe.

Benjamin compares the cameraman with the painter, but instead of immediately addressing the distinguishing features between the two, Benjamin compares a magician to a surgeon, and it is through this comparison that he makes clear how the painter and his technologically advanced counterpart, the cameraman, differ from one another. With regard to surgery, the magician is less involved than is the surgeon in that the former does not actually, as Benjamin states, penetrate the person. Benjamin explains, "The magician maintains the natural distance between the patient and himself [...] The surgeon does exactly the reverse; he greatly diminishes the distance between himself and the patient by penetrating into the patient's body" (1988b, 233). The magician never penetrates the person's body and never sees the inner organs, whereas the surgeon does. The surgeon *sees* the reality of the inner body, the magician does not. Something similar can be said about the painter and the cameraman. Benjamin elucidates: "Magician and surgeon compare to painter and cameraman. The painter maintains in his work a natural distance from *reality*, the cameraman penetrates deeply into its web" (233; emphasis added). Consequently, the works that these two artists produce differ in that the art of the painter is a "total one, that of the cameraman consists of multiple fragments which are assembled under a new law" (234). And like the cameraman editing a film in postproduction, a government can "edit" its historical narrative after the fact to serve whatever purpose, be it benign or malign, and unlike the average citizen or even esteemed academic, the government has the *authority* to proclaim its "final cut" of history as the *official* truth. Therein lies the analogy between film and governance, which is of great political consequence with regard to the acquisition and maintenance of power over others.

A salient example of this analogy of government as cameraman is the Japanese government's—specifically, former Prime Minister Shinzo Abe's—reliance on official documents as the sole representative of the truth concerning the issue of comfort women, a topic that will be discussed in more detail later in this chapter. During a budget committee meeting on March 5, 2007, Prime Minister Abe asserted that he did not see any evidence associating the Imperial

---

[2] Norimitsu Onishi (2007b) notes that Yasukuni Shrine's war museum "presents Japan as a liberator of Asia from Western powers."

military with the direct coercion of women into sex slavery. Abe based his argument on the premise that though there were official documents connecting the military to the establishment of sex brothels, there were no official documents linking the military to the direct coercion of women to serve in those brothels, stating, "There was no coercion such as kidnappings by the Japanese authorities. There is no reliable testimony that proves kidnapping" (*China Daily* 2007). Historian Yoshiaki Yoshimi counters Abe's assertion, explaining that the former prime minister's comments discounted the testimony of former comfort women and that history could not be constructed on the basis of "official documents" alone: "The fact is, if you can't use anything except official documents, history itself is impossible to elucidate," adding, "There are things that are never written in official documents [...] That they were forcibly recruited—that's the kind of thing that would have never been written in the first place" (Onishi 2007a). In the end, Abe's use of official documents alone to define history parallels the cameraman's decision to use certain scenes in the final cut of his film. "Unabridged" historical narrative is similar to *all* the footage shot for a film; it includes *everything*, ranging from official government documentation to eyewitness accounts and people's diary entries. Removal of anything from this total compilation constitutes "abridging," and quite possibly altering, the historical narrative. This alteration of the original narrative via the cutting of certain "scenes" thus transforms reality into the illusion of reality as the remaining "footage" is presented as the entirety of the narrative. Benjamin explains the illusion in film resulting from the editing process: "In the theater one is well aware of the place from which the play cannot immediately be detected as illusionary. There is no such place for the movie scene that is being shot. Its illusionary nature is that of the second degree, the result of cutting" (1988b, 233).

As the cameraman cuts scenes from the totality of the footage to construct his final version of a film, the Japanese government cuts "scenes" from the overall historical narrative—namely, all "unofficial" information—to create its version of history. Norimitsu Onishi (2007a) comments that the focus on only official documents as determining what or what isn't included in the historical narrative "has long been part of the government's strategy to control wartime history." As long as the Japanese government has within its disposal not only the ability to define "official history," but the ability to promulgate its version of history as the one and only truth, it has the power to maintain a degree of control over what the citizenry thinks about and how it thinks about it. And generally speaking, history is a subject area considered "boring" by students, most of whom—unless they have a keen interest in the subject matter, are history majors or aspire to become professional historians—probably limit their study of history to the confines of the classroom. Kazuo Ogura, president of the Japan Foundation and former Japanese ambassador to Vietnam, South Korea, and France; makes note of this lackadaisical attitude towards history among Japanese youth in an interview with the *Asahi Shimbun* (2008): "Perceptions of history naturally differ by country and person. [...] The problem is that young Japanese people have no interest in history and do not know much about it before discussing perceptions of history." Therefore, with knowledge of history limited to what is written in a textbook combined with the fact that the Japanese government has control over the content of said textbooks, the government also controls, to a certain extent, what students learn about history, molding these young people's points of view on many significant historical events.[3]

In summary, through the use of the media as a mode of disseminating information, a government, if it so chooses, can create reality and manufacture truth, as illustrated above with the case of the Abe administration's stance on acknowledging only official documents as truth, as the only available "footage" of history. And this creation of a reality is associated with translation. Take for example the death of civilians in war. These dead civilians can either be characterized as "casualties" or as "collateral damage," each word resulting in a completely different emotional reaction from the recipient of the news. The former characterization may induce emotions of sadness and anger, while the latter may induce no emotional reaction at all. Benjamin makes note of "a romantic grove" and "a melancholy lake" being called "a green grove" and "a blue lake," respectively (1999, 321; J51,4). What emotions do people feel when told of a *green* grove or *blue* lake as opposed to a *romantic* grove or *melancholy* lake? Like "collateral damage" and "casualties," the former terms used to describe the grove and the lake are likely to induce little, if any, emotional response.

Differences in emotional reactions can also occur when one substitutes the term "sex slave" with that of "comfort woman" or "prostitute." Assuming that a person does not have any prior knowledge of the Japanese comfort women issue (e.g., a Japanese student learning about World War II history for the very first time), by the government presenting comfort women as merely that—as prostitutes, with the implication that they were compensated for their services and voluntarily entered into such service—that person would probably not observe any ill-feelings toward the Japanese government; since prostitution was legal at the time, the government was not guilty of any heinous crime by employing sex workers. In contrast, if the government were to present these women as sex slaves, as victims who were sexually violated, this would likely entice strong emotions and a negative view of the

---

[3] Academic curricula in schools across Japan are approved and overseen by the Ministry of Education, Science, Sports, and Technology (MEXT for short). Thus, MEXT decides which textbooks are authorized for use in schools. MEXT oversees all stages of the textbook approval process, and those who are assigned to the textbook review committees are often of a conservative ideological slant (Hayes 2005, 192). Though "screening [of textbooks] is supposed to be free of political interference" (Onishi 2007b), ideology—a conservative one at that—nevertheless plays a role in education, in what knowledge a student ultimately acquires, or does not acquire.

actions committed by the Japanese government and military in the years prior to and during the Second World War. In short, reality is determined by the person (or group of people) who translates the events into the spoken or written word and presents them to the public for mass consumption. It is this power of manipulating words—modes of intention—to propagate a certain historical narrative or perception of reality that we shall explore next.

Translating 'Comfort Women': Modes and Objects of Intention

Benjamin's "The Task of the Translator" examines, as its title suggests, the work—and the responsibility—of the translator. What distinguishes a good translation from a bad translation? Benjamin answers these questions indirectly, articulating that "no translation would be possible if in its ultimate essence it strove for likeness to the original" (1988a, 73). In other words, with the exception of the difference in languages, a poor translation is one in which the translated version of a work is virtually identical to the original work. But since a translation is expected to convey the same content of the original work, but in another language, is the aforementioned statement then counterintuitive? The answer thus depends upon how one defines the purpose of translation.

In the beginning of his essay Benjamin states, "Translation is a mode" (1988a, 70). After an examination of the translatability of an original work, Benjamin then explicates the difference between "mode of intention" and "object of intention." Benjamin says that "kinship does not necessarily involve likeness" (74). The kinship that Benjamin refers to is that of languages, i.e., the relationship between the language being translated and the language employed for translation. Every language uses words to represent specific objects and concepts and unless they are closely related to one another, or one language borrows from the other, no two languages use the same word to represent the same idea. As Benjamin explains, "While all individual elements of foreign languages—words, sentences, structure—are mutually exclusive, these languages supplement one another in their intentions" (74). The distinction between *mode* of intention and *object* of intention is therefore of vital importance. A mode of intention is the word used to represent a certain object or concept, tangible or intangible. An object of intention, in contrast, is that to which a word refers. To illustrate the difference between mode and object, Benjamin uses the words *Brot* and *pain*. "The words *Brot* and *pain* 'intend' the same object," says Benjamin, "but the modes of this intention are not the same" (74). The German word *Brot* and the French word *pain* refer to—i.e., *intend*—'bread,' as we would refer to this object of intention in the English language. The intended object of *Brot*, *pain*, and *bread* is the same, but the modes, the words that denote, or the *names* given to, the object are clearly distinctive from one another. In the end, there are essentially two objects of intention: the object itself and the object perceived as a result of the particular mode used to refer to that object. Via the issue of Japanese comfort women, I will explain how differing modes of intention can actually shape the object of intention in the mind's eye.

Modes and Objects

The ongoing controversy surrounding comfort women, a euphemism used to describe the up to 200,000 women from Japan's colonies and territories across Asia (a majority of whom were Korean and Chinese) who were coerced or tricked into sexual slavery in brothels run by the Japanese military during World War II, stems from the Japanese government's reluctance to take full responsibility for, and its renewed desire to revise history concerning, the forced sexual slavery of these women.[4] Japanese school textbooks originally did not include references to the sex brothels set up and operated by the Imperial Army, but in 1997, a few years after the discovery of official documents directly linking the military to the creation of "comfort stations" and former Chief Cabinet Secretary Yohei Kono's 1993 statement acknowledging and apologizing for the Japanese military's involvement—directly or indirectly—in coercing women into sexual servitude, all school textbooks approved by the Japanese government included passages on comfort women.

In 2007, fourteen years after Japan formally acknowledged the existence of comfort women and the military's involvement and ten years after textbook revisions, a group in the Liberal Democratic Party (LDP), helmed by former Prime Minister Shinzo Abe and former Education Minister and current chairman of the "Fraternity of Lawmakers Who Are Concerned About the Future of Japan and History Education," Nariaki Nakayama, sought to, and literally succeeded in, rewriting history. Of the junior high school textbooks which were revised to include references to comfort women, only two now do (Onishi 2007a); the rest were re-revised to reflect the Abe administration's stance on the issue. Nakayama announced his satisfaction in removing any acknowledgement of comfort women from school textbooks: "Now few textbooks carry words like 'military comfort women' or that the women were 'forcibly taken' (to the frontline brothels). I think that's good" (Yoshida 2007). He went on to say that comfort women were not sex slaves, but prostitutes compensated for their services: "(Working at the brothels) was their commercial business. They were never sex slaves" (Yoshida 2007).

According to Nakayama, "In the first place, the term 'comfort women' didn't exist at the time (of World War II). What didn't exist is carried in his-

---

[4] Due to the destruction of official government and military documents pertaining to comfort women, estimates of the number of comfort women vary, ranging anywhere from 20,000 to 200,000. Two leading historians on the subject give opposing estimates. Chou University professor Yoshiaki Yoshimi claims that comfort women numbered between 50,000 and 200,000. Ikuhito Hata, a Nihon University lecturer who holds the position that there were no "sex slaves" (according to Hata, comfort women were more along the lines of professional prostitutes), on the other hand, estimates the number to be approximately 20,000 (Nakamura 2007).

tory textbooks. In other words, what is wrong is in the textbooks. Now I think it proper to see this (description) removed from the textbooks" (Jong 2007). This is "twisted logic," says *Chosun Ilbo* Tokyo correspondent Son-U Jong in a March 2007 editorial, asserting that just because the term "comfort women" did not exist in the past does not preclude the existence of comfort women. Benjamin may ask: Does the non-existence of a mode of intention infer the non-existence of an object of intention? As Benjamin explains, each language has its own mode of intention to refer to a particular object of intention. To reiterate, *pain* and *Brot* are two different words, but intend the same object: bread. For that matter, languages have synonyms, so one object of intention may be referred to by several different modes of intention, even in the same language. As a simple example, the object of intention 'woman' can be called *onna* or *josei* in Japanese—regardless of what word (mode of intention) is used, the object of intention is the same. Even if the words, *pain*, *Brot*, or *bread* did not exist in the lexicon of language, the object would still exist. Hence, as Jong argues, Nakayama's argument is flawed because even if the women coerced by the military or its middlemen into sexual service were not referred to as *ianfu* (comfort women) at the time—perhaps they were just referred to as *baishunfu* (prostitute) or some other term—they still existed and what happened to them, by their own testimonials, did occur.

Faithful Reproduction: The Task of the Translator

What exactly is the task of the translator? Benjamin articulates the following: "The task of the translator consists in finding that intended effect [*Intention*] upon the language into which he is translating which produces in it the echo of the original" (1988a, 76). This statement is an extension of what Benjamin mentioned earlier about a poor translation being a duplicate of the original work. The translator, as she or he is in the process of translating a written work, essentially has to take special care in the selection of words and phrases; the translator has to determine what modes of intention best convey the *spirit* of the original text. Take the Japanese word *sumimasen* for example, which means "I'm sorry," "excuse me" or "thank you" depending on context. *Kimochi warui* is another phrase with ambiguous meaning. This phrase, literally translated into English, means "feeling bad," but can convey this meaning or others (e.g., "That's disgusting" or "I feel sick") depending upon the context in which it is used. [5] A translator must therefore take context into

---

[5] In the final scene of *The End of Evangelion*, a Japanese animated film serving as the finale to the series "Neon Genesis Evangelion," the character Asuka utters the words "kimochi warui," which presented a difficulty to translators of the English version of the series for the context of the scene in which the lines are said was in itself quite ambiguous. Tiffany Grant, the voice actress who portrayed Asuka in ADV's English dub of "Evangelion" said the following regarding the ambiguous phrase: "The most widely circulated translation of the last line of EoE is 'I feel sick', but Amanda Winn Lee (voice of Rei Ayanami, who also directed EoE) said she asked several translators, and

account if she or he is to convey the essence of the original text. In short, the decision of which word or phrase captures the essence of the original text is dependent on the translator himself, and thus two different people tasked with translating an original work can and do produce different translations.[6]

Of finding this spirit of the original text, Benjamin elucidates: "Unlike a work of literature, translation does not find itself in the center of the language forest but on the outside facing the wooded ridge; it calls into it without entering, aiming at that single spot where the echo is able to give, in its own language, the reverberation of the work in the alien one" (1988a, 76). Drawing upon the analogy above, one can say that the translator stands outside the "forest of language" without entering it, "calls into it" and waits for a returning "echo" that resonates the mode of intention—the word or phrase in the translating language—that captures the true essence of the original work being translated. This having been said, the translator should not consider literalness to be the pith of a good translation. Literalness may in fact be detrimental to translation.

Benjamin mentions that fidelity and license are two considerations important to the task of the translator. Fidelity to the written word and the license of "faithful reproduction" (Benjamin 1988a, 78) often clash with one another. On the one hand, the translator must be veridical and precise regarding the words to be translated (fidelity), but must also have the freedom to use words and phrases that may not serve as a literal translation of the original work, though nevertheless communicate the *meaning* of the original work (license). Benjamin argues for license over fidelity as he says,

---

she felt 'disgusting' was the most accurate adaptation. You could say she is disgusted with/sick of the situation or with Shinji himself. [...] although (Eva creator) Hideaki Anno seems to change his mind frequently about what various things mean in Eva, Anno once said that Asuka's comment about feeling 'sick' was a reference to morning sickness" (2004).

[6] Of three prominent English translations of *The Tale of Genji*, each features a slightly different approach. According to Janice P. Nimura (2004), Arthur Waley, the first to translate *Genji* into English between the 1920s and 30s, "was more concerned with conveying the spirit than the letter of the original," while Edward Seidensticker, who produced a second translation in 1976, "stuck closer to the [original Japanese] text, conveying its sparseness as well as its stateliness and flashes of wry humor." The latest translation, written by Royall Tyler in 2002, "navigates a course between his predecessors," being "less baroque than Waley's, less brisk than Seidensticker's." Nimura then compares the translations of the opening line from *Genji by each of the three authors. This comparison* illustrates that there is no one "true" translation, that the interpretation of an original text truly is dependent on the perspective of the translator. Waley: "At the Court of an Emperor (he lived it matters not when) there was among the many gentlewomen of the Wardrobe and Chamber one, who though she was not of very high rank was favored far beyond all the rest." Seidensticker: "In a certain reign there was a lady not of the first rank whom the emperor loved more than any of the others." Tyler: "In a certain reign (whose can it have been?) someone of no very great rank, among all His Majesty's Consorts and Intimates, enjoyed exceptional favor."

> Fidelity in the translation of individual words can almost never fully reproduce the meaning they have in the original. For sense in its poetic significance is not limited to meaning, but derives from the connotations conveyed by the word chosen to express it. We say of words that they have emotional connotations. A literal rendering of the syntax completely demolishes the theory of reproduction of meaning and is a direct threat to comprehensibility. (78)

Benjamin cites Hölderlin's translations of Sophocles as examples of the flaws and "monstrousness" of such direct, literal translations, true in meaning to the original *text* (i.e., the individual words), but void of the original emotional conveyance of the writings (78). A translation is like a jigsaw puzzle that, once put together, resembles the original image depicted in the puzzle, but is imperfect due to the visible cracks in the image; nonetheless, this imperfection may very well represent the perfection of the translation.

As stated above, the primary issues of the debate on comfort women are the extent of the Imperial military's involvement in setting up and/or running sex brothels and the coercion of these women to work in the brothels. This debate aside, Benjamin would probably simplify the issue by examining the term "comfort women" itself. Does this term capture the essence of what these women were? Is "comfort women" a proper *translation* of who these women were? To state another way, is "comfort women," as a mode of intention, a "faithful reproduction" of the objects of intention that have been labeled "comfort women"? Conservatives like Shinzo Abe and historian Ikuhito Hata would say yes. Others like historians Yoshiaki Yoshimi and Hirofumi Hayashi would disagree; Yoshimi himself prefers the term "sex slave" to "comfort woman" (Nakamura 2007).[7] To those who view comfort women as victims, coerced by the military into providing sexual service against their will, the term "comfort women" is a euphemism that does not serve as an adequate description of what these women were. For instance, Hata, who strictly uses the term *ianfu* (comfort women) in addressing these women, "refuses to say sex slaves because he claims the women, and

---

[7] On April 17, 2007, Yoshimi and Hirofumi Hayashi, Professor of Politics at Kanto Gakuin University, announced the discovery of documentation from the 1946-48 International Military Tribunal for the Far East (Tokyo War Crimes Tribunal) pertaining to the *Tokeitai* (Imperial Navy's Military Police) seizing women for the sole purpose of serving in the military brothels. One of the documents, dated March 13, 1946, quotes Shuichi Hayashi of the Imperial Navy stating that women in Indonesia were arrested on false charges so that they could be taken to comfort stations to work: "I admit to have slapped these women with the flat of my hand; I also ordered them to undress [...] I do not think these women were actually punishable, but their arrest ... was only a pretext to put them in a brothel" (Tabuchi 2007). Yoshimi, at a news conference announcing the new evidence, urged the Japanese government to take full responsibility for the military's actions: "The evidence is clear that the system of 'comfort stations' was set up, maintained and enlarged by the Japanese military itself [...] The government must issue a clearer message that responsibility lies with the military ... Japan should also accept legal responsibility, and offer individual compensation" (Tabuchi 2007).

according to historians, girls, were not forced into the frontline brothels. Hata claims the women were trading sex for money" (Nakamura 2007). Yoshimi, in contrast, "explicitly refers to them as sex slaves" (Nakamura 2007), as these women were forced into sexual servitude and imprisoned against their will.

Based on his stance on fidelity and license (again, he argues for license over fidelity), Benjamin would likely argue against using the term "comfort women," reasoning that "comfort women" exhibits fidelity to the original object of intention, but fails as a faithful reproduction of said object. To explicate, fidelity to the object is retained via "comfort woman" for it is essentially a literal translation of the object: the object of intention is a *woman* whose purpose was to provide *comfort* to soldiers; hence, she was a *comfort woman*. But this literal term does not capture the *essence* of what this woman was: a sex slave. To recapitulate, Benjamin mentions that words have "emotional connotations," which are at times necessary to convey the "poetic significance" of the original object of intention (1988a, 78); the words "comfort women" eradicate any of these emotional connotations, and their literalness is ultimately a "direct threat" to the original object's comprehensibility (78).[8]

In the end, it is the terminology that frames the issue at hand. If one were to distinguish between the two terms, denying the existence of *comfort women* is one thing, while denying the existence of *sex slaves* is another. If Shinzo Abe, for instance, were to maintain that comfort women did not exist, he would be treading on thin ice since the evidence and testimony available does not support his premise: that there were no comfort women. But on the other hand, if Abe were to say that sex slaves did not exist, according to his definition of a sex slave—in comparison to his definition of a comfort woman (i.e., a professional paid prostitute)—he would technically be correct. In the end, the term "comfort woman" and the term "sex slave," though each may refer to the same *actual* object of intention, can arguably result in two very different *perceptions* of the object of intention. "Comfort woman" generally intends "professional paid prostitute," a woman that is compensated for her services and provides these services willingly. In contrast, "sex slave" intends a *victim*, someone forced into sexual servitude against her will with no (or limited) rights and privileges. As Akemi Nakamura (2007) points out, "Hata figured the situation [of comfort women] was similar to prostitutes at regular brothels, which were legal those days. However, Yoshimi says the sex slaves were that by definition—they did not have freedom to leave or refuse sex with soldiers." Additionally, if a comfort woman is defined

---

[8] With regard to fidelity and license of the translator, perhaps U.S. Ambassador to Japan, J. Thomas Schieffer, presents a *faithful* mode of intention for comfort women: "victims of rape." In a press conference, Ambassador Schieffer said of comfort women: "I think they were coerced to engage in prostitution. That means that they were raped by the Japanese military at that point in time. [...] I think that happened, and I think it was a regrettable, terrible thing that it happened. I think the events speak for themselves" (*Chosun Ilbo* 2007c).

as a professional prostitute, then liability can be shifted away from the Japanese government, for prostitution and the operation of brothels were legal at the time (Nakamura 2007). If a comfort woman is defined as a sex slave, then, according to a 1998 UN report regarding the Japanese government's legal liability for the operation of comfort stations during World War II, the Japanese government is guilty of and liable for institutionalized slavery (the establishment of comfort stations), war crimes and crimes against humanity (coerced prostitution; enslavement and rape of women on such a large scale) (McDougall 1998). The locations where the comfort women were held can also be said to be euphemized with the term "comfort station"—the same report uses the wording "rape centre" to describe the same object of intention. One editorial on comfort women makes note of words as modes of intention: "These were not commercial brothels. Force, explicit and implicit, was used in recruiting these women. What went on in them was serial rape, not prostitution" (*New York Times* 2007).

The Definition of 'Coercion'

In his 1993 statement, Yohei Kono makes three references to the coercion of comfort women:

> The Government study has revealed that in many cases they were recruited against their own will, through coaxing *coercion*, etc.
>
> They lived in misery at comfort stations under a *coercive* atmosphere.
>
> The Korean Peninsula was under Japanese rule in those days, and their recruitment, transfer, control, etc., were conducted generally against their will, through coaxing, coercion, etc. (emphases added)

In a public statement he made on March 1, 2007, Shinzo Abe picked apart the definition of the word "coercion." Referencing the 1993 Kono statement, Abe said that there was "no evidence of coercion" (*Chosun Ilbo* 2007b), that the Imperial military's responsibility "must be premised on a change in what constitutes the definition of coercion" (*Chosun Ilbo* 2007f). In October 2006, Abe said that women were coerced into becoming comfort women "in a narrow sense" of the word (*Chosun Ilbo* 2007f). Thus, women were either coerced or they weren't, depending on how one defines "coercion." Jong (2007) summarizes Abe's logic as follows:

> According to Abe, there is coercion in a strict sense, such as when military officers raided civilian homes and abducted women. And there is coercion in a broad sense, such as when businessmen, under the instruction of Japanese troops, forced women to serve as sex slaves. Abe claimed that since there is no evidence to prove the strict sort of coercion, it must follow that Japanese soldiers never coerced women into sex slavery.

Definitions matter a great deal, especially if it is the government doing the defining. An editorial featured in the *Chosun Ilbo* (2007e) notes that prior to becoming prime minister, Shinzo Abe was a leading member of a faction within the LDP that sought to revise Japanese wartime history, and at the time, publicly referred to the military's role in the comfort women system as a "fabrication." The commentator also makes the following observation: "… the comments he made as a lawmaker and those he makes as a prime minister have very different ramifications. His comments as a prime minister represent the official line of the Japanese government" (2007e).

This is the power of translation—the power of interpretation—and it is a power that those in positions of authority use to influence and even manipulate the thoughts and perceptions of the general public. Benjamin cautions us of this, for it is through the restructuring of facts and the creation and propagation of "official truths" that allow a society's "power elite" to influence the thoughts and perceptions of the masses. "The Task of the Translator," in brief, presents the use of translation/interpretation as a means for governments to define reality and influence public perception, but this is not the only control mechanism at the disposal of governments. Benjamin's "Theses on the Philosophy of History" explores one other method of thought control, one that, unlike translation, leaves reality behind.

Of Puppets and Puppeteers: The Power of Myth

"Theses on the Philosophy of History" opens with a brief commentary on an automaton playing a game of chess on a tabletop. "A system of mirrors created the illusion that this table was transparent from all sides. Actually, a little hunchback who was an expert chess player sat inside and guided the puppet's hand by means of strings" (Benjamin 1988c, 253). Like puppets, the masses are strung along to believe a certain point of view. Governments can construct history, and this history may not be complete, highlighting only those events that are perceived as necessary to serve a certain purpose, whatever that purpose may be. For instance, political leaders may shape history in such a way as to get the public to think in a certain manner or to legitimize their control over the populace. And one way of defining history is through the use of mythology. The leaders of years past and the present day, no matter what nation, society or culture, have employed myth at one time or another, with some myths more farfetched than others. In the case of Japan, for instance, according to ancient Japanese mythology, not only the Imperial family, but the islands that make up Japan itself, are descended from deities. Being descended from gods, the Emperor himself can be considered a god or at least god-like, which was the *official* view perpetuated by the Japanese government to the people prior to Japan's surrender after the Second World War.

In relation to puppetry and controlling the masses, the Japanese Emperor being portrayed as a deity is not unlike a character portrayed by a *film* actor who is subject to the manipulations of a "higher authority"—namely, the director. Benjamin explains the difference between stage actor and film actor: "The stage actor identifies himself with the character of his role.

The film actor very often is denied this opportunity. His creation is by no means all of a piece; it is composed of many separate performances" (1988c, 230). It is the cameraman/director, not the actor himself, who defines the character through cuts and edits made after the conclusion of filming. In that respect, the actor is like a puppet manipulated by the cameraman, just like the puppet playing chess is manipulated by the hunchback. As the chess-playing puppet is akin to the public, the film-acting "puppet" is akin to the Japanese Emperor as he was defined in the centuries preceding the end of World War II. This was particularly the case during the Meiji Restoration of 1868, when the Tokugawa Shogunate was removed and the Imperial family restored to power. But even then, the Emperor was really nothing more than a symbol of national unity and an "embodiment of political authority" (Hayes 2005, 18). True power rested with a "small elite" (18) that could utilize the Emperor's name to maintain the legitimacy of their authority over the state.

When all is said and done, the ability to not only shape reality, but to get a majority of the public to *believe* it, gives a government great power. Fantastical myths aside, modern-day myths are more realistic in tone. Ultimately, in retroactively shaping official history and defining moments after the fact, like a cameraman editing scenes and defining a character during post-production, a government can decide what events to include and what to exclude, as well as who to label as heroes and who to denote as villains. As Benjamin annunciates:

> To articulate the past historically does not mean to recognize it "the way it really was" (Ranke). It means to seize hold of a memory as it flashes up at a moment of danger. Historical materialism wishes to retain that image of the past which unexpectedly appears to man singled out by history at a moment of danger. The danger affects both the content of the tradition and its receivers. The same threat hangs over both: that of becoming a tool of the ruling classes. (1988c, 255)

Perceptions obviously differ from person to person, from group to group, and history may in a sense be subjective, and by that token, it does matter whose view of historical events is proclaimed as "official" to the general public. Again, returning to the subject of Japanese wartime history, it is the Ministry of Education that has oversight of the content that goes into textbooks that are read by students nationwide. Who are the people that sit on MEXT's textbook review committees? What are their views? Many people obviously have contrasting views of Japanese history and the extent of the atrocities (whether the actions are to be labeled "atrocities" is a debate of its own!) committed by the Japanese government prior to and during World War II, but in the end, the beliefs of the review committee members do matter more than others, since it is their perceptions of history that are disseminated as the mainstream view of history across Japan.

The committee members' ideology ultimately plays a role in determining what is factual, i.e., factual in the eyes of the government. For example, was Imperial Japan a "liberator" or a "conqueror"? Were Japan's actions "aggression" or "advances"? Were comfort women "professional prostitutes" or "sex slaves"? Thus, history is not as objective as it would appear to be. In terms of history, the object of intention—the historical event itself—may be objective, but the mode of intention—the narrative used to describe the event—is subjective. Benjamin warns of such great power being vested in the hands of a fascist government, but he does acknowledge that all is not lost. In order to engage the fascist state, the masses must exploit the same tactics:

> The tradition of the oppressed teaches us that the "state of emergency" in which we live is not the exception but the rule. We must attain to a conception of history that is in keeping with this insight. Then we shall clearly realize that it is our task to bring about a real state of emergency, and this will improve our position in the struggle against Fascism. One reason why Fascism has a chance is that in the name of progress its opponents treat it as a *historical norm*. (Benjamin 1988c, 257; emphasis added)

In brief, in order to effectively defy and overthrow the Fascist state, opponents of fascism must reject the "historical norm," the history created by the fascists themselves to perpetuate their rule, and establish their own history. To harness the power of history, one must link relevant historical events to the present day or, as Benjamin would say, "blast [them] out of the continuum of history" (1988c, 261).

Into the Phantasmagoria

In "The Work of Art in the Age of Mechanical Reproduction," "The Task of the Translator," and "Theses on the Philosophy of History," Walter Benjamin explores the perception of reality and how to make use of words and definitions to shape the truth. This is clearly illustrated by the subjective nature of history, which exists in many different versions, depending on who is telling the story. History is a temporal narrative taken from fragments and constructed after the fact, and may or may not be true. The truth is open to translation and constructed from various fragments, put together in a way that serves to benefit those in power. Apropos politics, with the advent of mechanical reproduction came a new age in which fantasy and reality seem to converge, allowing a government, if it so chooses, to utilize illusion to effectively shock, awe, and manipulate the masses.

In a debate between Shinzo Abe and Democratic Party of Japan (DPJ) member Toshio Ogawa during a Diet budget committee hearing in March 2007, Ogawa disagreed with Abe's claim that there was no evidence directly associating the military with the coercion of comfort women, stating, "Unless Japan offers an apology ... I am afraid the international community will think Japan has not learned the lesson on human rights or from the war, which Japan started" (*Chosun Ilbo* 2007a). Abe rebutted, making note of Japan's six decades of "highly appreciated" actions and asserting that Ogawa was ignoring these sixty years of Japanese history. Ogawa responded: "I am disregarding Japan's past?" (*Chosun Ilbo* 2007a).[9] The same question could be posed to Abe and others who deny Japan's involvement in the coercion of comfort women and the operation of sex brothels during the war in spite of evidence stating the contrary.

For the sake of argument, let us say that during the Second World War the Japanese Imperial military did indeed force women into sexual servitude. Now consider the following hypothetical: It is now 2040 or some other future date and all hard evidence linking the military to sex slavery has been destroyed and all former comfort women, soldiers, and any other eyewitnesses are long dead. The Japanese government then proclaims something along the lines of what it is arguing now: "Comfort women? Never existed. Military coercion of such women? Never happened." In addition, any references to comfort women have already been erased from school textbooks. Perhaps a few people—historians and descendants of comfort women, for instance—again, with no hard and direct evidence at their disposal, say that comfort women did exist, basing their claims on stories told to them by those who are already deceased. The government then responds that these people are misinformed, having been told, and are now telling, myths and lies. As a result, a kind of "future amnesia" is created via the revision (i.e., erasure) of history.[10]

In the end, the above hypothetical is not so far off from what is happening today as neo-conservatives in Japan's national government are, as Ogawa says, "disregarding Japan's past," and employing propaganda to do it. Jong (2007) explains:

> To provoke the Pacific War, Japan killed Zhang Zuolin in a bomb attack and blew up the Liutiaohu Railway. It also destroyed the proof of its war crimes and blamed China for what happened. The Japanese fell for the propaganda and swore vengeance. Considering these moves, it seems only logical that Japan would have destroyed any and all records of comfort women. And now Japan is ramping up the propaganda again, claiming that comfort women never existed because there is no evidence to prove otherwise.

For centuries, propaganda has been a tool all governments and ruling bod-

---

[9] An excerpt of the words exchanged between Abe and Ogawa is presented in the *Chosun Ilbo* article, "Abe Digs in Heels Over Comfort Women" (2007a).

[10] Gerard P. Heather. "Political Theory as Action" (lecture, San Francisco State University, San Francisco, CA, April 15, 2005).

ies employed for promulgating the truth from their point of view. Whatever the issue, assuming a lack of evidence supporting either side, a debate can boil down to a case of 'he said, she said' between the government, in which the public trust has been vested, and the group of people who disagree with the government's "official truth." Who are the citizenry more likely to believe? Who or what should they believe? Who should be acknowledged as having the "proper view," to use former Japanese Prime Minister Tomiichi Murayama's words, of the past?[11] For most people, reality is whatever the government says it is: "It rests with the prelates, the nobles, the great officers of the state to be the repositories and guardians of the saving truths, to teach nations what is bad and what is good, what is true and what is false ..." (Benjamin 1999, 344; J65,1).

"I don't see myself as taking a political role. I see myself as exposing the truth. [...] Doing this may not always be in the interests of Japan. But that's the role of the historian," says Yoshiaki Yoshimi (*China Daily* 2007). But in a world clouded by phantasmagoria where lies become truths and truths become lies, what is *truth*?

---

[11] In rebuttal to former Prime Minister Abe's comments denying the extent of the military's role in the comfort women issue, former Prime Minister Tomiichi Murayama (1994-96) is quoted as saying, "It is an indisputable fact that the Japanese military set up and operated 'comfort centers' and the Japanese government is responsible. [...] It is almost pointless to argue whether the Japanese military played a role in forcibly recruiting comfort women," adding that Abe and others do not have a "proper view" of history (*Chosun Ilbo* 2007d).

CHAPTER 2

Tokyo Phantasmagoria

"Behold our great cities under the fog of tobacco smoke that envelops them, thoroughly sodden by alcohol, infused with morphine: it is there that humanity comes unhinged. Rest assured that this source breeds more epileptics, idiots, and assassins than poets," quotes Walter Benjamin in Convolute J, a section of *The Arcades Project* devoted to the writings of Charles Baudelaire (1999, 250; J13,1). This depressing and lackadaisical image is not often one that comes to mind when a person thinks of Europe's "great cities" of the 19th century. What of today's great cities? Are they too breeding grounds for epileptics, idiots and assassins? Behind the beautiful architecture and technological advancements of grand metropolises are characteristics that great civilizations seek to hide under a veil of fog that is the phantasmagoria.

The collection of literary quotations, notes, and reflections that were eventually complied and published as *The Arcades Project* was Benjamin's attempt to critically analyze the "commodification" of life in 19th-century Europe—specifically, in Paris—exploring how materialism and consumerism had shaped European society. Though Benjamin wrote about the 19th century, much of what he had written applies to the consumer society of the present day. Drawing directly from *The Arcades Project* as well as commentary from Susan Buck-Morss's *The Dialectics of Seeing*, this chapter will elucidate how Benjamin's mammoth thirteen-year undertaking is still relevant in today's world. Of *The Arcades Project*, Buck-Morss says, "Ostensibly a social and cultural history of Paris in the nineteenth century, it is in fact intended to provide a political education for Benjamin's own generation" (1991, 48). As this chapter will illustrate, Benjamin's analysis is not restricted to his generation; it applies to our generation as well. Contemporary myths of commodity capitalism will be explored by focusing on Japan, with particular attention to the city of Tokyo, Japan's archetype of the modern (or *post*-modern) city, and how the modern world has become more engulfed in commodity and commercialization since Benjamin's time.

The focus of Chapter 2 is to elucidate how a society's retrogression is actually disguised as progress, and how consumerism and commodity fetishism are leading to the commodification of humanity—both mind and body—elaborating upon Japan's education system and sex industry. But first, the phantasmagoria of modern Japanese life, the series of beliefs conveyed unto the people by commodity capitalism, will be exposed by showing that modernity has yet to usher in a utopian society without social or economic differential; specifically, modernity has yet to eradicate prejudice and poverty. Benjamin mentions a simple example of phantasmagoria—that of the flâneur: "The phantasmagoria of the flâneur: to read from faces the profession, the ancestry, the character" (1999, 429; M6,6). To infer who or what a person is merely by her or his outer appearance is not unlike the adage of judging a book by its cover. In like manner, one should not judge Tokyo by its majestic surface image, but delve deeper to discover what lies beneath. This discussion of the "Tokyo phantasmagoria" and humanity's commodification will serve as a prelude to the topic of the final chapter of this discourse on Japanese social and political life: The application of art and allegory to "demolish the harmonious façade of the world" (329; J55a3) and reveal the "grotesqueness" of reality.

Materialism and the Humanization of Commodity

As "temples of commodity capital" (Benjamin 1999, 83; A2,2), the arcades captured the attention of those that strolled pass them, enrapturing people in awe and wonder and "housed the last dinosaur of Europe, the consumer" (Buck-Morss 1991, 65). This dinosaur left behind descendants—and their stomping grounds stretch beyond the European continent. Consumerism is very much alive today, but in an entirely different form than that which was the focus of *The Arcades Project*. Gone are the arcades of old, but they have been replaced by the department store. In short, Paris was a phantasmagoria, a word Benjamin used to describe the city as "a magic-lantern show of optical illusions" (81). The same term can be used to describe almost every major city in the world, including Tokyo.

Materialism as a Value System

Tokyo is an exemplification of materialism made manifest in a country of "hyperconsumers" (Ritzer 2005, 221). Walking down the streets of Tokyo, one will see young and old people alike sporting designer and brand-name apparel and accessories, from Burberry scarves to Louis Vuitton purses; a Tokyoite is essentially a "strolling commodity" (Benjamin 1999, 367; J79,4). In terms of wealth and lifestyle, today's Japanese youth and young adults enjoy greater luxury and financial wellbeing than previous generations. Many single women and men financially depend on their parents' income to support them and are thus able to take pleasure in quite affluent lifestyles. Of women in particular, Masahiro Yamada explains, "The reason why the standard of living of young women is high, in spite of the fact that their incomes are low and they have a high unemployment rate, is that they depend economically on their parents" (1998, 36). Of women who do work, in 1996 over 80 percent of these women continued to live with their parents (36). In brief, it is quite common for a woman's parents to financially provide for her until she marries. For those women who are married and *not* working, many are able to enjoy a significantly greater amount of recreational time than their *salarīman* ("salaryman") husbands and, as Midori Yajima explains, their "search for happiness and the consumption this involves seem to have replaced frugality as the highest virtue" (1992, 65).

Yajima affirms Yamada's analysis, stating, "Japanese women [...] live a life of material plenty and convenience that would have been beyond the widest imaginings of their ancestors" (1992, 63). In earlier times, Japanese women were quite frugal in their everyday lives. Yajima gives the example of women passing down their kimonos to the next generation. In the present day, it seems that the culture of frugality has given way to a culture of fashion, consumerism, and the pursuit of pleasure. Fashion magazines are displayed on every newsstand, presenting the latest fashion trends that young people are not at all hesitant to try out. One merely has to walk down a busy street or through a subway station in Tokyo, which are analogous to fashion-show runways with men and women showcasing the clothing and accessories featured in *Gainer*, *Men's Knuckle*, *NonNo*, *Pinky*, and other Japanese fashion magazines.

"The arcades are a center of commerce in luxury items. In fitting them out, art enters the service of the merchant," notes Benjamin of the 19th-century Paris arcades in his 1935 exposé, "Paris, the Capital of the Nineteenth Century" (1999, 3). Like the arcades of yesteryear, Tokyo's shopping districts are the heart of commodity goods and art, the latter of which, in the form of the ubiquitous advertisement, serves the purpose of the seller. Billboards and large video screens advertising the latest "must have" product are everywhere, and many *baito-kun* (part-time workers) themselves have become human billboards, stationed outside their respective store or business handing out flyers or other promotional items advertising their store's product or service.[12] To get the attention of the "flâneuring" consumer and potential customer, some of these workers are equipped with loud speakers and/or dressed in colorful costumes ranging from French maid outfits to *Pokémon* characters. These walking billboards are the archetype of "the integration of the human body into advertising" (186; G7,4). Though Benjamin's flâneur is not a consumer, but a stroller on the outside looking into the phantasmagoria of commodity capitalism, the potential of the flâneur becoming consumers—even becoming the merchandise itself—parallels the following description by Benjamin mentioned early on in Convolute A:

> Trade and traffic are the two components of the street. Now, in the arcades the second of these has effectively died out: the traffic there is rudimentary. The arcade is a street of lascivious commerce only; it is wholly adapted to arousing desires. Because in this street the juices slow to a standstill, the commodity proliferates along the margins and enters into fantastic combinations, like the tissue in tumors.—The flâneur sabotages the traffic. Moreover, he is no buyer. He is merchandise. (1999, 42; A3a,7)

Unarguably, the desire to acquire money and the commodities that money can buy is one shared by many who are raised in a commodity-capitalist society, but in Japan, materialism has become not only a pastime but a value system. "'God be praised, and my shops too.' The arcades as temples of commodity capital," notes Benjamin (1999, 37; A2,2). In another religious reference, Benjamin states: "On Baudelaire's 'religious intoxication of great cities': the department store are temples consecrated to this intoxication" (61; A13). Commodity takes on a divine quality as materialism is worshipped as a god and the department store, like the arcades of old, becomes akin to a place of worship: a temple. According to psychologist David Matsumoto, "In my observations, one value that many Japanese have adopted is materialism […] Quite frankly, the youth of Japan have come to treasure money and what it can buy. […] What strikes me as peculiar is the meaning that this search for material goods has acquired in Japan" (2002, 125). In simple terms, capitalism is a combination of object and fantasy.[13] This is taken to the extreme as human qualities are attributed to these inanimate objects. Capitalism manipulates us as a commodity product becomes the recipient of human emotions like love. Take Valentine's Day for example. A day for couples to celebrate their love and af-

---

[12] These promotional items include pens and fans (the latter of which are most prevalent during the summer), but the most common are tissue packets with an ad printed either on the packet itself or on a piece of paper inside the packet. I accumulated quite a collection of these tissue packs during my one year in Tokyo, having acquired three on the evening of my arrival.

[13] James R. Martel. "Susan Buck-Morss's *The Dialectics of Seeing*" (seminar, San Francisco State University, San Francisco, CA, April 10, 2006).

fection for one another, some critics say Valentine's Day is nothing more than an occasion for the greeting-card, confectionary, and floral industries to make a profit.[14] As Benjamin cites from *Les Murailles revolutionnaires de 1848*, "certain manufacturers profit from certain occasions" (Benjamin 1999, 85; G7,1). Valentine's Day just so happens to be one of those occasions. The giving of the commodity product is a symbol of the person's love and by extension, the product itself may be loved. The phrase "Diamonds are a girl's best friend" is a perfect illustration of this humanization of commodity. Saying that a diamond can be a "best friend" attributes the human quality of friendship to an inanimate commodity item, the diamond. This begs the question: how does love for commodity affect one's love for fellow human beings?

Obsession of the Otaku:
The World of Commodity Fetishism in Modern Tokyo

An extreme example of love for—and the sex appeal of—commodity can be seen through the Japanese *otaku* subculture. *Otaku*, a word originally used to refer to another person's house or home (although the word is still used in this manner), also refers to a person who has an obsession with a certain interest, particularly in *anime* or *manga*, Japanese animation or comics, respectively.[15] The stereotypical *otaku*, who can best be described as a "pathological-techno-fetishist-with-social-deficit" (Gibson 2003, 95), can often be seen roaming the Akihabara district of Tokyo purchasing *anime* or *manga*-related merchandise, including action figures and model kits (mostly of the female characters), and video games.

Of the video games available to the Japanese consumer, one genre that is popular among *otaku* is the dating simulation game. In these "dating sims," which play like a "choose your adventure" book in which you are given choices to make throughout the game and whatever choice you make advances the story and affects your future choices and outcomes, the objective of the game is

---

[14] Japan actually has two days of the year to celebrate love (and for certain industries to make a little extra profit): Valentine's Day and White Day, which is exactly a month later on March 14th. In Japan, Valentine's Day is the day women give chocolates (or other gifts) to men and the men reciprocate on White Day. Though there are several theories about the creation of White Day, one of them explains that it was devised by a marshmallow confectioner sometime in the 1960s for the primary intent of selling marshmallows (hence the name White Day) by appealing to the Japanese value of repaying social obligations; specifically, of the man giving a return gift to the woman who gave him chocolates on Valentine's Day (Japan-guide.com 2000).

[15] The terms *anime* and *manga* mean animation and comic book, respectively. In Japan, these terms refer to animation and comic books in general, regardless of where they are produced. Outside of Japan, *anime* and *manga* commonly take on a definition that is limited in scope, denoting animation and comics of Japanese origin. It is this narrower definition that will be applied when the words *anime* and *manga* are used in this essay.

to court a potential lover (or multiple lovers) and may or may not include explicit sex scenes.[16] Benjamin, while discussing the phantasmagoric nature of world exhibitions, notes the potential isolation resulting from a person's connection with commodity: "They open a phantasmagoria which a person enters in order to be distracted. The entertainment industry makes this easier by elevating the person to the level of the commodity. He surrenders to its manipulations while enjoying his alienation from himself and others" (1999, 7). As a person identifies with a commodity item, she or he faces the possibility of *not* identifying with other people. One is then "elevated," as Benjamin states, to the level of the commodity and alienates herself or himself from the company of other living beings, finding solace in having a "relationship" with the commodity item. Dr. Shinya Iwamuro of the Japanese Association for the Development of Community Medicine's Health Promotion Research Center echoes Benjamin's concern as he references young men with a deep interest in dating sims, expressing his apprehension about their fears of being in real human-to-human relationships: "They don't want to get hurt [by being dumped by human girls], so they never advance past the 2D world" (Anime News Network 2007).

In addition to stores selling *anime*-related merchandise, another popular location for *otaku* to "hang out" is the maid café, in which customers are served by waitresses wearing French maid outfits. Other such *cosplay* (short for "costume play") cafés and shops exist in Tokyo and throughout Japan, where customers can be serviced by school girls, policewomen, race queens, or "bunny girls." One may go so far as to say that these women are relegated to the role of a human dress-up doll, representing the merging of the woman with commodity. Benjamin mentions this combination of woman and commodity, saying, "Under the dominion of the commodity fetish, the sex appeal of the woman is more or less tinged with the appeal of the commodity" (1999, 345; J65a,6). For *otaku* and non-*otaku* alike that frequent these cafés, it may not necessarily be the woman's physical beauty that they are attracted to, but the costume that the woman wears. Libido is aroused, thus, by the woman's clothing, and not by the woman herself, signifying the "assimilation of the living to dead matter" (354; J71,3). Of fashion and commodity fetish, Benjamin states:

> Fashion prescribes the ritual according to which the commodity fetish demands to be worshipped. [...] Fashion stands in opposition to the organic. It couples the living body to the inorganic world. To the living, it defends the rights of the corpse. The fetishism that succumbs to the sex appeal of the inorganic is its vital nerve. The cult of the commodity presses such fetishism into its service. (8)

---

[16] Though most dating sims are geared toward heterosexual male gamers in which the gamer assumes the role of a male protagonist attempting to get a girlfriend, games which feature a female protagonist searching for a boyfriend and games primarily focused on homosexual relationships are also available.

These maid cafés are therefore paragons of commodity fetishism in Japanese society. In this age of commodity fetish, what arouses a heterosexual male's sexual desire more—a woman's naked body or the lingerie she wears? That having been said, it should also be noted that several stores in Tokyo sell not just clothing for "everyday wear" but costumes as well—from the standard school girl uniforms to costumes worn by *anime* or video game characters—for a woman to purchase 365 days of the year (the costume racks are not limited to Halloween season in Japan) for whatever purpose, but perhaps for the intent of enhancing her sex appeal to her boyfriend or husband. In this case, it is not art imitating life, but life imitating art, as women become the real-life versions of the 2D cartoon and video game characters of men's fantasies. "It is [...] the clothing business that furnishes the prototype of the modern man and woman. ... Mannequins become the model for imitation, and the soul becomes the image of the body" (Benjamin 1999, 78; B8,4).

Commodity fetishism, thus, imbues the inorganic with sexuality, something that is fundamentally associated with the organic human body: "Not seldom in the dream, however, there are swelling breasts that, like the earth, are all appareled in woods and rocks, and gazes have sent their life to the bottom of glassy lakes that slumber in the valleys. These landscapes are traversed by paths which lead sexuality into the world of the inorganic" (Benjamin 1999, 70; B3,8). As the woman finds companionship in the diamond, a man finds sexual gratification in a school girl's sailor suit or maid's apron.

The Phantasmagoria of Fashion and Japan's New Comfort Women

One can say that fashion is a self-contained phantasmagoria of its own for it creates an illusion. As a woman applies make-up to her face and covers herself with clothing, she essentially conceals her true self and indulges in the image of the woman she has become via her make-up and dress. Strip all of the outer layers away, and you are left with just the woman, with the underlying, and conceivably grim and depressing, reality: "If a woman of taste, while undressing at night, should find herself constituted in reality as she has pretended to be during the day, I like to think she'd be discovered next morning drowned in her own tears" (Benjamin 1999, 63; B1,6). It is this reality that one escapes as she or he lives in the phantasmagoria; and some do so by choice because they do not want to accept the dreary truth of this underlying reality. Take the case of an elderly woman with her gray hair concealed with dye, her blemishes with mascara, her varicose veins with dark stockings or a long skirt. She escapes death, or at least attempts to; and that is the power—the allure—of fashion: "Here fashion has opened the business of dialectical exchange between woman and ware—between carnal pleasure and the corpse. [...] And that is why she changes so quickly; she titillates death and is already something different, something new, as he casts about to crush her. For a hundred years she holds

41

her own against him" (62-63; B1,4). In the final analysis, this discourse on fashion is a metaphor for the concealment of the social problems many societies such as Japan face today—instead of looking directly at these problems, people seek to cover them up.

And what of a woman, young or old, who opts for cosmetic surgery as an attempt to attain "perfection"? This *seikei bijin* or "plastic beauty" becomes a human mannequin, an amalgam of the organic and inorganic, of life and commodity. Human perfection is thus found not in other humans, but in the inorganic commodity, which Benjamin says is exemplified by the wax figure, in which "the surface area, complexion, and coloration of the human being are all rendered with such perfect and unsurpassable exactitude that this reproduction of human appearance is outdone" (1999, 409; L2a,1). As was done above with fashion and social ailments, we can find a metaphor between beauty and politics: by erasing her natural flaws—by "revising" her original form—through surgery, a woman is akin to a government that erases its "blemishes" and "scars" of the past through revisions of historical rhetoric. If we were to link this discourse on fashion and phantasmagoria—i.e., the concealment of the problems not only of the past, but of the present—to a situation in modern Japan, one that would illustrate this association vividly is that of human trafficking.

It may not be a country known for the 'export' of people for exploitation, but Japan does its fair share of 'importing' women to meet the demands of the country's sex industry. Women who are trafficked into Japan are not unlike the comfort women of the past. Many are tricked, lied to, coerced, or even kidnapped, and forced into prostitution, pornography, or other forms of sexual service. Many women who end up in sex servitude in Japan often come on the false promises of high-wage jobs as waitresses or factory workers, but "nearly all" end up with enormous amounts of debt and are forced to provide sexual services to repay those debts (Dinan 2000).

Human trafficking is "the fastest growing criminal industry in the world" because of the low risk of prosecution and the ability to make a lot of profit in very little time (Polaris Project 2004). Generally speaking, if there is a lack of demand or no demand at all for a certain commodity, production and selling of the commodity is likely to decrease or cease completely as the producer/seller will make little or no profit from such a commodity. Therefore, since the demand for "human resources" for the purposes of forced labor and prostitution is particularly high, there is a significant number of humans trafficked to regions of the world where demand is great. The Polaris Project (2004), a non-governmental organization that combats and raises awareness about human trafficking, explains, "Traffickers profit from consumer demand for cheap labor and goods and for sex by exploiting vulnerable populations for slave-like labor or sexual exploitation." And in Japan, with its ever booming sex entertainment industry, demand is high.

An editorial in the *Chosun Ilbo* (2006) iterates a classic phrase on history, while urging Japan to teach history accurately: "We study and teach history lest we are condemned to repeat it. If it genuinely wants stability, peace and common prosperity in Asia, Japan must teach history as it happened to present and future generations in Asia." As can be seen from the continued trafficking of humans and prostitution in Japan, phenomena that can be attributed to the *demand* of such sexual services, it looks like history is repeating itself. "The impression of the old-fashioned can arise only where, in a certain way, reference is made to the most topical. If the beginnings of modern architecture to some extent lie in the arcades, their antiquated effect on the present generation has exactly the same significance as the antiquated effect of a father on his son," says Benjamin (1999, 69; B3,6), referring to how the past influences the present. Like fashion and architecture, current political and social dilemmas often have a basis in the past. Because Japan is unable to acknowledge the atrocities of the past, by extension, it appears not to acknowledge the atrocities committed in the present. Women trafficked into Japan for the purpose of sex slavery are "comfort women" in all but name. In short, the exploitation of women in Japan's plethora of sex brothels is an example of the "eternal return of the same" (71; B4,1).[17] Human Rights Watch (1994) succinctly states the reality of human commodification in the modern world: "It is long past time for the international community to realize that women and children in many parts of the world are being sold on the international market like any other commodity." On a similar note, Benjamin states the following in respect to one's affection for the prostitute, who is essentially a commodified human: "Love for the prostitute is the apotheosis of empathy with commodity" (1999, 375; J85,2).

---

[17] Japan has an active and wide-ranging sex industry, with some services considered legal or outside the bounds of legal interference. These businesses, which have "remained outside the purview" of the definition of illegal prostitution, are actually regulated under the Law on Control and Improvement of Amusement Businesses (Dinan 2000). These "sex entertainment businesses" include, but are not limited to, "image clubs," where women dress in costumes and role play, acting out the fantasies of their customers, and typically provide oral sex (think maid cafés but with a sex component); "soaplands," where women lather their customers with soap (hence, the name "soapland"), bathe them, and often proceed with sexual intercourse; "telephone clubs" (*telekura*), where a customer pays to be called by a woman to arrange a meeting for a "compensated date" (*enjo kōsai*) with the general implication that the meeting will conclude with sexual intercourse; and "hostess clubs" (*kyabakura*), well-known fixtures of Japan's red-light districts, most notably the Kabuki-cho district of Tokyo's Shinjuku ward, where customers go to drink with and be entertained by the clubs' female employees. The primary duty of the hostess is to provide company and entertainment for the man, chatting with him, listening to his rants, or doing whatever her customer pleases, but supposedly there are restrictions on what a hostess can do with a customer. Hostesses are not prostitutes per se; sexual intercourse is not in their "official" job descriptions, but it is not uncommon for some hostesses—willingly or unwillingly—to give a customer sexual services "on the side."

The Fantasy World of Capitalism and Japan's "Internal Colony"

Japan is home to amusement parks such as Tokyo Disneyland and its next-door neighbor Disney Sea, Universal Studios, and Sanrio Puroland. These theme parks, like any other, can be seen as representative (albeit an exaggerated representative) of the fantasy of modern life. One goes to an amusement park to be amused—to have fun and enjoy the entertainment value of the park's fantasy world. The park is basically a large playground in which one can forget the stress and anxiety of school and work. Caught up in the fantasy, one may suspend reality for the time being and forget that there is actually a person inside that Donald Duck or Hello Kitty costume, and see these costumed people as the characters that they portray. This suspension of reality occurs outside of the theme park as well, in our daily lives, as we see the riches and splendor of modern life, but ignore or refuse to acknowledge its imperfections. For example, how many people take for granted the luxuries that they have—no matter how big or small, from the clothing they buy to the food they consume—but then turn a blind eye as they pass a vagrant on the curb as though these impoverished men and women do not exist in the commodity world in which they live? Just as we see an amusement park worker's costume and not the person inside, throughout our everyday lives, we are constantly exposed to the surface reality and rarely the underlying reality of life. In sum, a commodity-capitalist society is one big amusement park, a "world of mist" (Benjamin 1999, 101; D1,1) where fantasy and myth clouds the senses.

The efficacy of myth is that it can make people forget about what is actually happening around them. Benjamin begins Convolute E, "Haussmannization, Barricade Fighting," with the following from Franz Böhle's *Theater-Catechismus*:

> The flowery realm of decorations,
> The charm of landscape, of architecture,
> And all the effect of scenery rest
> Solely on the law of perspective.
> (1999, 120)

Böhle shows how beautiful structures and elegant designs—the flowery realm of decorations—can influence perceptions of our surroundings. It was Baron Haussmann who employed the power of aesthetics to its fullest. In 1852, Haussmann, hired by Napoleon III to "modernize" the city of Paris, utilized illusory tactics in the form of "urban 'renewal' projects" (Buck-Morss 1991, 89). The projects of this "Haussmannization" of Paris were designed to help create a social utopia by rearranging and redesigning the city, either by moving buildings and public structures or by creating new ones (89). What these projects actually accomplished was the creation of the *façade* of a utopian society, or as Buck-Morss describes it, "the illusion of social equality," for "class antagonisms were [...] covered up, not eliminated" (89).

"The reconstruction of the city ..., by obliging the workers to find lodgings in outlying *arrondissments*, has dissolved the bonds of neighborhood that previously united them with the bourgeoisie," quotes Benjamin from Levasseur's *Histoire des classes ouvrières et de l'industrie en France* (1999, 123; E2,1). It appears that people from different socio-economic spectrums were intermingled at one point in time, but after Haussmann, were segregated from one another, the upper class living in one area of the city and the lower class in another. This is similar to the divisions in any modern city, as "the most scandalous alleys ... disappear to the accompaniment of lavish self-glorification by the bourgeoisie ..., but—they reappear at once somewhere else, often in the immediate neighborhood" (145; E12,1). In brief, poverty is not dealt with; it is simply moved elsewhere. Even the utopian-like 'megapolis' of Tokyo is home to an area that most Tokyoites may not have even heard of, a district so unlike the glitz and glamour of the rest of the city that it does not even appear on maps of Tokyo.[18]

San'ya is a neighborhood in Tokyo's Taito ward, which is home to many so-called "day laborers," comprised primarily of impoverished elderly men with limited formal education who move from job to job (mostly in construction) without a sense of security on what their next source of income will be. For centuries, since the Tokugawa Era, low-ranking workers and their families have resided in San'ya, sometimes referred to as "the other Japan" or "Japan's internal colony." "Anywhere you go, you'll find a negative side and a positive side," says Jin Kigoshi, a craftsman who has lived and worked in San'ya for over a decade, explaining the existence of San'ya in Tokyo's post-modern, futuristic metropolis (Harvey 2004). San'ya is a place that many Japanese do not talk about, either because of prejudice and bigotry linked to the past[19] or just because of sheer ignorance or shame of San'ya's existence.[20] Benjamin gives the following quote by Roger Caillois about the power of phantasmagoria over human perception:

---

[18] Ty Harvey (2004) notes that "San'ya is not in guidebooks or on subway maps—or any other maps at all."

[19] During the Tokugawa Era, also known as the Edo Period (1603-1867), the area that is present-day San'ya was home to the *burakumin*, a minority group that primarily worked in occupations that were considered "unclean" at the time (e.g., butchering and preparing leather). These people were discriminated against, and this discrimination continues to this very day against people residing in San'ya, regardless of whether they are directly descended from the *burakumin*.

[20] Harvey (2004) recalls his first time in San'ya, noting that the residents themselves seemed hesitant to acknowledge San'ya's existence: "On my first trip to San'ya, I saw the vacant buildings, the crowds of homeless, and even though I wasn't sure that I was there, I was sure that I was close. So I asked a local shopkeeper where I could find it, and he pointed vaguely in one direction. I asked someone else, and she pointed in another. Later, I found out from a bartender down the street that I'd been in San'ya all along."

"It seems reasonable to say that there exists ... a phantasmagorical representation of Paris (and, more generally, of the big city) with such power over the imagination that the question of its accuracy would never be posed in practice—a representation created entirely by the book, yet so widespread as to make up ... part of the collective mental atmosphere." (1999, 439; M12,1)

Like Paris, the "accuracy" of Tokyo is rarely questioned by its inhabitants; what they see *is* their reality. Excluding those who live or work there, how many Tokyoites have actually *seen* San'ya first hand, let alone know of its existence? When one thinks of Tokyo, what comes to mind are probably the department stores, skyscrapers, billboards, technology, the nightlife, and of course the daily commute of millions of people in the crowded streets and subway stations. What about homeless people, decayed buildings and empty roads? Certainly these are not images one thinks of when she or he is asked to picture Tokyo in her or his mind's eye. But these images are nevertheless the reality that is San'ya. Ty Harvey (2004) makes the following observation:

> The balance is there between Shibuya, in Tokyo's southwest, with its high-rise department stores and Luis [sic] Vuitton billboards and herds of giggling schoolgirls in a neon lap of luxury, and San'ya, in Tokyo's northeast, with its half-vacant blocks, littered streets and a thousand homeless men lined up in the park for their one free meal of the day. [...] In short: San'ya is Tokyo's slum.

Benjamin says that "every city is beautiful to me (from outside its borders)" (1999, 458; N1,8). Acknowledging the existence of San'ya is to go beyond the borders of the "beautiful" city of Tokyo. Like Haussmann's Paris, Tokyo hides its socio-economic differentials behind aesthetically pleasing structures and displays. The homeless, not only in San'ya, but throughout Japan, are ultimately ignored and forgotten, seen as nothing more than insignificant blights in a society that values material wealth. Benjamin presents the following quote by Friedrich Engels regarding homeless people and how the rest of society leaves them alone to fend for themselves: "And those who cannot pay for ... a shelter? They sleep wherever they find a place, in passages, arcades, in corners where the police and the owners leave them undisturbed" (44; A4a,2). If success were to be defined as the founding of a utopian society, Haussmann failed because he focused strictly on aesthetics, phantasmagoric in substance, for the renovations merely hid class divisions and avoided addressing the underlying social problems plaguing the city: class conflict. Though there have been no proletarian uprisings in modern Japan, the potential for a clash—violent or otherwise—is existent. Kigoshi states ominously: "If San'ya were gone [...] and if these beaten down, good-hearted people were to lose their place to run to, sooner or later they'd destroy Japan" (Harvey 2004). Like Paris after Haussmannization, modern-day Tokyo is a phantasmagoria, a dream world where many social problems are ignored or even forgotten.

Commodification of Humanity

Just as human attributes have been applied to commodity items, humans in turn exhibit qualities inherent to commodities. This "commodification" of human beings has both a physical and mental aspect. Of the former aspect, prostitution and the sex industry in general play significant roles, which is especially the case in Japan since it has a flourishing sex entertainment industry comprised of several adult video distributors and various types of brothels. As for the latter aspect, knowledge has become a commodity in and of itself, available for purchase by students, the consumers of knowledge. The prevalence in Japanese education to 'teach to the test' (i.e., university entrance examinations) have resulted in the ubiquity of *juku*, "cram schools" that specifically train students in preparation for taking entrance exams. In "Cultural Influences on the Construction of Knowledge in Japanese Higher Education," Ken Kempner and Misao Makino postulate where Japanese education appears to be headed, alluding to knowledge becoming a product of "packaged education" that can be sold to anyone who can pay for it:

> *Jukus*, and the form of specific, packaged education they provide are so essential to Japanese preparatory education that some politicians propose ending public support of education at the elementary and secondary levels, allowing the *juku*-type education to replace public schools. Privatisation of schools would create a system for the sole purpose of passing college entrance examinations. (2004, 189)

In the end, living in a society where the acquisition of wealth and commodity goods is held in high regard, where the amount of money one has is essentially equated to one's probability of survival, humans are susceptible to commodifying themselves as they are at the same time humanizing the objects they so dearly desire.

Commodification of the Human Body:
The Quest for Commodity, the "Evolution" into Commodity

As mentioned above in the section on the phantasmagoria of fashion, the sex industry in Japan is a high-profit business, and not all women who become sex workers are coerced into it. Many women, as well as young girls, *voluntarily* enter the business. Why do they do it? The short answer: Sex sells—and some women want 'a piece of the action.' In the end, the root cause of this is Japan's commodity culture and how women have been commodified as a result of it. As is often said, prostitution is "the world's oldest profession," though Benjamin goes so far as to call prostitution "a precursor of commodity capitalism" (1999, 348; J67a,1). J.T. Brown (2003), in an opinion piece on *enjo kōsai* or "compensated dating," ponders the question of why a native Japanese female living in a nation "where everyone has food in their bellies and ample material toys to play with already" would willingly become a prostitute.

47

Delinquent teens from broken homes would certainly be the stereotypical type of female that goes into prostitution of her own free will. But that presumption is merely that: a stereotype. Many participants in *enjo kōsai* are actually from families of a middle-class economic background (Kingston 2005, 27). Brown (2004) affirms, stating that "plenty of girls and women from very ordinary circumstances and upstanding families have been found doing this" and asserts that the reasons are generally either the ease of making a large sum of money within a short period of time or a need for male affection (especially from older men), which can be attributed to the lack of such affection from the father.[21] Though Benjamin does not explicitly say so, he does allude to the absence of parents in the following quotation from *Lettres à sa mere*: "Where is the mama who will give *Les Fleurs du mal* as a present to her children? And where is the papa?" (1999, 313; J46a,2). Just as parents were absent in their children's lives in the Paris of the 1800s (the above excerpt from *Lettres à sa mere* was written in 1861), they continue to be absent in present-day Tokyo. Thus, regarding the lack of fatherly or, generally speaking, parental support and affection in a child's life, these are yet other examples of the eternal return of the same, which Benjamin makes note of several times in *The Arcades Project*, primarily in Convolute D. It is not implausible to argue that the failure of parents to be involved in their children's lives is partly attributed to the (over)time parents spend at work, not necessarily because they want to but because they have to. In brief, the need to labor and to make end's meet—to survive—is the incurable disease of modernity and the commodity-capitalist lifestyle that comes with modernity. The need to labor, in the end, not only consumes one's life but becomes one's life, with familial bonds being one of the expenses—or casualties, to use another mode of intention—of financial security and the acquisition of wealth.[22]

Regarding *enjo kōsai*, though there may be some women who are forced into such service, there are many who participate by choice. The women who engage in *enjo kōsai* are a diverse group, consisting of single and even married middle-aged women, but tend to comprise of young women and girls. Of child prostitution cas-

---

[21] Matsumoto presents the results of a 1998 survey administered by the *Asahi Shimbun*, which asked respondents to rate the causes of *enjo kōsai*. Most people (34 percent) felt the major cause of *enjo kōsai* was parents' inability to "discipline their children" (2002, 125). It is interesting to note that the second highest response, chosen by 30 percent of those surveyed, was "society overemphasizes the desire for goods" (125).

[22] Alice Yamada (1996), whose friends have engaged in *enjo kōsai*, blames parents for not giving children the emotional support that they require during their adolescent years and states the following about the dire consequences of improving economic strength at the expense of family ties: "Japan is experiencing a degeneration of morals and ethics after the entire nation had devoted itself to becoming a strong economic power. Children were left to fend for themselves because the adults were too busy making money and as a result, the teenagers now are living lives as they please."

es involving *enjo kōsai* arranged via "dating sites," the National Police Agency of Japan states that in 93.8 percent of the cases (as of 2002), it was the girl that initiated first contact (Kioka 2003). On a related note, this evident promiscuity of women is not a recent phenomenon and should not be treated as such; again, it is yet another illustration of Benjamin's motif of eternal recurrence. Of women's virtue in 19th-century Paris, Benjamin cites the following quote from Comte Horace de Viel-Castel's *1883 Mémoires sur le règne de Napoleon III:* "As for the virtue of women, I have but one response to make to those who would ask me about this: it strongly resembles the curtains in theaters, for their petticoats rise each evening three times rather than once" (1999, 491; O1a,1). Though in 21st-century Tokyo, these "curtains" require less effort to raise since the modern "petticoats" that young female Tokyoites tend to wear are commonly of a much shorter variety.

"The rank of the species is in direct proportion to its resemblance to the human being," quotes Benjamin from Toussenel's *L'Esprit des bêtes* (1999, 195; G12a,1). In addition to the purpose of making money, the mindset that women are not equal in status to men, or that women are sub-human, serves to justify to the trafficker or pimp that it is okay for him to collect and sell these women as sexual *objects*. Women who willingly prostitute themselves are no different; most, if not all, are also inspired by the profit motive. Benjamin presents the following excerpt from Egon Friedell's *Kulturgeschichte der grande dame,* which notes how money has become the object of love and sexuality and also makes a reference to prostitution as the modern-day "erotic ideal": "Making money becomes the object of an almost sensual fervor, and love becomes a financial concern. In the age of French Romanticism, the erotic ideal was the working girl who gives herself; now it is the tart that sells herself" (75; B6a,2).

For *enjo kōsai*, rates generally range between 30,000 to 50,000 yen "for a two hour tryst" (Brown 2003). This amounts to approximately 600,000 to 1,000,000 yen a month if a girl works one "tryst" a day for twenty days during the month. In comparison, according to the Japanese Ministry of Internal Affairs and Communications (2007), the average starting salary in 2006 for women who recently graduated from four-year universities was approximately 191,000 yen per month. Additionally, prostitutes who decide to go into "private business" have essentially eliminated the middleman and are thus able to keep all their earnings; hence, they are not only the products, but the sellers—they sell themselves. Some *enjo kōsai* working girls do not always ask for monetary compensation, but for commodity items: "Alternative forms of payment can be directly taking a girl to a department store and paying for the designer shoes or jacket she picks out, or bringing along such a gift to the time and place of an appointment" (Brown 2003). The desire for commodity goods as a reason for selling one's body—this has been the case in the past and continues to be in the present day. Of the women-turned-prostitutes of the 19th century, Benjamin quotes F.F.A. Béraud:

"I have finally succeeded in discovering the true principle of this increase in prostitution … at … certain times of the year. […] girls like to give and receive presents or to offer beautiful bouquets; they also want a new dress for themselves, or a hat in the newest fashion, and, lacking the necessary pecuniary means, … they turn for some days to prostitution to acquire such means." (1999, 501; O6,2)

The teenage girls and young women of 21st-century Japan who go into prostitution are no different. Not having the "necessary pecuniary means" of purchasing the objects of their desire, they allow themselves to be bought to acquire such means.

Yasumasa Kioka (2003) of the National Police Agency presents examples of messages posted on the websites of virtual *telekura*. The typical male postings include:

"Will pay 50,000 yen maximum for Enjo Kosai."

"Awaiting messages from high school girls short in pocket money."

Of the female postings, these include:

"Send me a mail if you can meet me Sunday for 30,000 yen."

"I'm a teenager in high school. Let's get together if you're somebody who can give me pocket money. Send profile."

"I want to try real sex! I'm bored with playing with myself. Somebody with lots of money please make me feel good."

I have chosen these five postings in particular to illustrate the profit motive exhibited by both males and females who participate in *enjo kōsai*. As can be seen in the messages, men appeal to the girls' desire to make money by offering to provide them with dearly desired "pocket money," while the girls directly request money in their postings. Even the girl that is "bored with playing" alone specifically requests a person "with lots of money" to be her companion. Jeff Kingston also notes the desire for money and commodity goods as a personal motivation for girls who participate in *enjo kōsai*, explaining that "indulging in the fantasies of these middle-aged 'infants' [i.e., the clients] and ripping them off are portrayed as part of the game […] the girls want discretionary income to purchase expensive designer clothes and accessories and pay their *keitai* (cellular phone) bills" (2005, 27). Ultimately, prostitution makes women into commodity items to be bought and sold, and is the greatest example of how humans have been commodified—either by others or by their own free will. In their desire for commodities, humans themselves have become commodities.

Commodification of the Human Mind:
The Purpose of Education–of Life–in a Commodity-capitalist Society

Like the human body via prostitution, the human mind, too, via the indoc-trination of an education system that creates "corporate soldiers," has been subject to the "commodification" process. For all intents and purposes, the general goal of schooling is to impart knowledge, but is education limited to merely the delivery of knowledge or is there a deeper purpose? Japanese law-makers of the past noted the importance of education in the development of a well-rounded citizenry and the creation of a better society. An excerpt from the Japanese Fundamental Education Law of 1947 states: "Education shall aim at the full development of personality, at rearing a people, sound in mind and body, who love truth and justice, esteem individual values, respect labor, have a deep sense of responsibility, and are imbued with an independent spirit as builders of a peaceful state and society" (Kempner and Makino 1993, 188). With regard to Japanese higher education, the purpose of universities has been to train corporate managers and government bureaucrats to be "builders of a peaceful state and society." Is the Japanese education system fulfilling this purpose today?

Walter Benjamin himself was a learned man, a philosopher and scholar who valued scholarship and sought to impart knowledge through his writings. One could argue that Benjamin would support a strong and efficient education sys-tem; not one that produces mindless drones, but one that produces creative and innovative thinkers, especially those that do not fear challenging authority and the status quo if and when necessary. If Benjamin were alive today, he would probably be appalled at the current state of education in Japan, for the Japa-nese education system, as he would likely see it, is not a breeding ground for intellectuals who will strive to create a "beautiful society," but a factory manu-facturing cogs—or in Benjamin's words, automatons—whose primary objec-tives are not necessarily the creation of a better society but merely the ascent up the social hierarchy. "Within this educational structure," explain Kempner and Makino, "knowledge is a commodity only to improve one's chances to gain entrance to the next level" (1993, 198).

Students' creativity and critical thinking skills are inhibited by rote mem-orization of facts, of "pre-packaged" knowledge, necessary for the purpose of passing entrance exams. Hideyuki Negishi explains that Japanese students "have never had the time or room to attempt anything but the instructions of-fered in manuals and, hence, no room for exercising their creativity" (Doyon 2001, 448). This inhibition continues during one's college years either as a re-sult of the classes themselves not being academically rigorous or due to the student's own lack of interest and motivation to engage in the coursework. Kempner and Makino affirm this, stating, "The higher education system is criticised for its lack of creativity, parochialism, rigour and for its low quality of instruction" (1993, 185). Many Japanese universities are considered "leisure lands," and even corporate managers are complaining about the low quality of today's college graduates (Doyon 2001, 444). If anything, today's Japanese youth are, generally speaking, the product of commodity-capitalism: They are a "prosperity generation" or *shinjinrui* ("new human species"), with a leisurely, carefree lifestyle, and unlike their parents and grandparents, are prone to spending rather than saving (449).

The purpose of education in Japan today, though hidden under the government's phantasmagoric rhetoric of developing one's personality with the purpose of making positive contributions to society, is actually an economical pursuit. Knowledge, as Kempner and Makino suggest, has become a pre-packaged commodity. Some of this knowledge is free, available to all students through their primary and secondary education. Other knowledge "packages," sold by *juku*, private schools, and all institutions of higher education, are available at a premium to those who are willing and able to pay. In short, the focus of education is more about monetary gain than it is about passing on knowledge. The selling of pre-packaged knowledge and teaching to the test have led to a pervasiveness of *juku* and *eikaiwa* (private English language schools) throughout Japan. The profit motive—Benjamin would see this as the primary purpose of the education-corporate complex that has become a fixture not only in Japan, but in nations across the world. For example, private companies that offer courses geared toward standardized test preparation—be it the SAT, GRE, or GMAT, to name a few—are part of a highly profitable industry in the United States that are comparable to the Japanese *juku* and *eikaiwa*.

Though it was previously stated that the student is the consumer, it appears that they are not the only ones doing the consuming. In order for schools, be it a university or *eikaiwa*, to endure, they need "human resources" in the form of students. As living beings consume food to survive, academic institutions consume students to survive. Lack of a sufficient number of students ultimately results in the school's shutdown—in death. It is the student that is being consumed by "commodity education," just as the laborer is consumed by the means of production: "It is now no longer the laborer that employs the means of production, but the means of production that employ the laborer. Instead of being consumed by him as material elements of his productive activity, they consume him as the ferment necessary to their own life process" (Benjamin 1999, 195; G12a,3).

Making a profit may be the purpose of the education system for school administrators and education-oriented private corporations, but education also has an economic profit-making component for students. What is the ultimate goal of students who strive hard academically (at least during their pre-college years)? Is it to one day build a better society or maybe go further and attempt to build a better world? For some students, that may be a desire they wish to fulfill upon graduation and becoming *shakaijin* or full-fledged members of society.[23] For others, perhaps for the majority, it is to gain the qualifications to get a decent job, make money, and survive in this "struggle for existence in modern society" (Benjamin 1999, 436; M10,3). In short, unless a person is significantly affluent and does not need to retain a job to be financially secure, one lives to work, and works to live. In modern Japan, a *shakaijin* is merely an automaton that provides her or his intellectual or physical services in return for a wage—in return for the chance to survive. When all is said and done, it may not be farfetched to compare the Japanese *shakaijin* to the prostitute—broadly defined.

The American Heritage Dictionary (2004) offers the following two definitions for "prostitute":

1. One who solicits and accepts payment for sex acts.
2. One who sells one's abilities, talent, or name for an unworthy purpose.

The first definition can be said to be "prostitute" narrowly defined, while the second is "prostitute" broadly defined. Benjamin is known to redefine words, applying his own definitions to existing words or modes of intention.[24] Thus, I shall now do something similar with the word "prostitute," though I will not create an entirely new and radically different definition for the term. What I will do is broaden the definition further, not limiting it to sex acts or unworthy purposes: "One who sells one's abilities, talent, or name for monetary gain." By this broader definition, anyone who works is a prostitute. Benjamin presents the prostitute as "seller and sold in one" (1999, 10). When one engages in a paid occupation, like those that sell their bodies for money, isn't any employee, whatever the profession may be, seller and sold in one? A worker sells one's self, one's knowledge or skills, in return for monetary compensation. Does that not make one a prostitute by the broadest definition of the word? Benjamin would probably say so, as he does associate prostitution with work and vice versa: "Prostitution can lay claim to being considered 'work' the moment work becomes prostitution. In fact, *lorette* was the first to carry out a radical renunciation of the costume lover. She already arranges to be paid for her time; from there, it is only a short distance to those who demand 'wages'" (348; J67,5).[25]

As we attribute prostitution to any monetarily compensated profession, we go beyond the "beautiful" borders of phantasmagoric reality, encountering a

---

[23] The word *shakaijin* is used to refer to a working adult, but the combination of the two words making up the term, *shakai* (society) and *jin* (person), can also denote another meaning for the word: a member of society. What is a member of society? Related to the purpose of education is the purpose of the *shakaijin*. What is her or his role or purpose in society? Merely to work and make end's meet or to accomplish something more *substantive*?

[24] James R. Martel. "Walter Benjamin's 'Theses on the Philosophy of History'" (seminar, San Francisco State University, San Francisco, CA, February 20, 2006).

[25] Howard Eiland and Kevin McLaughlin, the translators of the English version of *The Arcades Project*, define *lorettes* as "ladies of easy virtue" (Benjamin 1999, 978).

different way of viewing and perceiving life in a commodity-capitalist society. Take for example, the description of a working-class woman in Victor Hallays-Dabot's play *Jenny the Worker*: "*Jenny the Worker* brings to life one of the most terrible afflictions of the social organism: the daughter of the working class ... constrained to sacrifice her virtue for her family, and to sell herself ... in order to provide bread for her loved ones" (Benjamin 1999, 706; a4a,5). Although the excerpt is from a work published in 1871, it is germane in any epoch, and in any civilization. Jenny the Worker can easily be replaced by Stephanie the Worker or Sayumi the Worker. Jenny could be a factory worker. She could be a prostitute. On the basis of that brief excerpt alone, we cannot pinpoint her exact profession. Benjamin leaves this ambiguous because the description can be applied to anyone, working any job; any working-class person can "sacrifice her virtue" and "sell herself" so that she can ensure her survival as well as the survival of her "loved ones." Baudelaire himself took a broad view of prostitution, as he "occasionally saw the publication of poems as a form of prostitution" (330; J56a,3). Benjamin presents an excerpt from the following poem by Baudelaire, addressed to Sarah, his first mistress:

> Though to get some shoes she sold her soul,
> The good Lord would laugh if with this wretch
> I struck a haughty pose like some Tartuff,
> I who sell my thought and would be an author.
> (266; J21,2)

Just as the prostitute sells herself—her soul?—for a pair of shoes, a writer sells his thoughts to become an author, and in essence, prostitutes his mind as the harlot prostitutes her body.

Like narrowly-defined prostitutes (sex workers) who go from john to john, broadly-defined prostitutes (workers of any profession) are akin to "expendable assets," whose services are utilized for the benefit and profit of whatever company or organization she or he works for and is discarded (i.e., fired or laid off) when her or his services are no longer required, and forced to look for a new job, a new source of income, in order to survive. This is now the case in Japan, where lifetime employment is no longer a luxury that workers have. Again, without money, many of us, unable to afford food, clothing, or shelter, would be dead. And assuming lack of affluence, in order to get money, one has to *work* for it—prostitute one's intelligence or manual labor—in exchange for monetary compensation. This is the purpose, not only of work, but of *life*, a life that Japanese education prepares its students for. In the end, education in Japan has not necessarily succeeded in developing well-rounded *shakaijin*, but has essentially succeeded in producing, as Hiroshi Fukunaga elucidates, "good *kigyō senshi*, or corporate soldiers for Japan, Inc." (Kempner and Makino 1993, 185). In that respect, the Japanese education system has failed to achieve the aims set out in the 1947 Japanese Fundamental Education Law. In a society where money and the material goods that money can buy are quintessential to survival, the "love [of] truth and justice, esteem [for] individual values" and "a deep sense of responsibility" (188) are perhaps "luxuries" too expensive to afford. As Jean Jacques Rousseau (2006) once wrote, "It is too difficult to think nobly when we think for a livelihood."

Where is Love?

Chapter 1 offered a discourse on the political power of interpretation and the ability of government to manipulate the truth to create and propagate an illusory reality or mythic history. This chapter delved further into the illusory nature of commodity-capitalism: the phantasmagoria. We have explored the phantasmagoria of Tokyo, Japan's post-modern city, revealing the underlying reality of Japan's impoverished classes, the failure of the education system to create *shakaijin* (as opposed to *kigyō senshi*), and materialism as the overarching value system for Japanese youth. In *Oeuvres*, Baudelaire describes a future world that is far from the utopia of people's dreams:

> Of the coming society, it is said that "nothing in the sanguinary, blasphemous, or unnatural dreams of the utopians can be compared to what will actually happen. ... Rulers will be compelled, in order to maintain their position and create a semblance of order, to resort to methods that would appall present-day mankind, hardened as it is. ... Justice—if, in this fortunate epoch, any justice can still exist—will forbid the existence of citizens who are unable to make a fortune. ... Those times are perhaps quite close at hand. Who knows whether they are not here already—whether it is not simply the coarsening of our natures that keeps us from noticing what sort of atmosphere we already breathe?" (Benjamin 1999, 315; J47a,3)

This world, as Baudelaire alludes, may already be the one humankind is currently living in, but has failed to realize it. The illusory phantasmagoria of the wonders of modern life, along with the adoption of materialism as a system of morals and values, have "coarsened" the nature of many Japanese—in particular, of Japanese youth—to the point that they are unable to see the grim nature of the reality in which they live. For example, their government officials, in order to "maintain their position"—in the Japanese case, their honor and pride—are able to resort to "appalling methods" of rewriting history to subdue or erase the atrocities their predecessors had committed, while the impoverished, "unable to make a fortune," are not permitted to "exist" (to reiterate, San'ya, home of Tokyo's underclass, is not listed on any official maps).

Overall, in *The Arcades Project*, Benjamin describes the commodification of human life in 19th-century France, something that is a reality in modern-day Japan. As commodities have been humanized, humans have concurrently been commodified. In the same way that items are "marked with fixed prices" (Benjamin 1999, 58; A12,1), certain types of humans are valued at different prices. For example, a corporate lawyer is generally "priced" more (i.e., paid more) than an elementary school teacher, who in turn is generally "priced" more than a custodian. Benjamin quotes from the first chapter of Karl Marx's *Das Kapital*:

> "A commodity appears, at first sight, a very trivial thing and easily understood. Its analysis shows that in reality it is a very queer thing, abounding in metaphysical subtleties and theological niceties. So far as a value in use, there is nothing mysterious about it. ... The form of wood is altered by making a table out of it; nevertheless, this table remains wood, an ordinary material thing. As soon as it steps forth as commodity, however, it is transformed into a material immaterial thing." (196-97; G13a,2)

Assuming that all humans are created equal, like the piece of wood that is altered through the production process, after going through the education system to become *kigyō senshi*, each person is still human, but is nevertheless transformed as a result of her or his education (or lack thereof)—one becomes a doctor, police officer, secretary, cashier, etc.—and has a price affixed, denoting the person's "worth."

"My pussy is worth $600," says a girl who participates in *enjo kōsai* (Yamada 1996). So exactly how much is a human being worth? In the end, the love of commodity has taken precedence over—and for some, has replaced—the love of human life. As Setsuko Inoue laments, "We lost the most important thing to find value in something invisible, like love" (Fujiwara 1998).

CHAPTER 3

Of Art and Allegory

The first two chapters dealt, for the most part, with the illusory nature of reality—of what is *perceived* as reality—in modern Japan. These illusory realities have been depicted in two forms, one being applied retroactively by government via translations and interpretations of "official" historical narrative and the other the phantasmagoria, an illusion—an image of the ideal world—propagated not solely by those in power, but by the masses themselves. The final chapter of this discourse on Walter Benjamin and commodity-capitalism in contemporary Japan will focus on two of Benjamin's tools to dispel the mist of phantasmagoria: the political use of art media by the citizenry—what will be referred to as "the people's camera" or "anti-propaganda"—and the application of allegory.

Political media can take various forms, ranging from artwork and films to newspaper editorials, all of which provide political and social commentary not unlike what Benjamin did with his writings. The potential for media to be used as a means of enlightening the populace and enticing them to political action is alluded to in "The Work of Art in the Age of Mechanical Reproduction." In the preface of the essay, Benjamin states the following of the emerging forms of art and print reproduction: "They are [...] useful for the formulation of revolutionary demands in the politics of art" (1988b, 218). He goes on to say that media such as film can be modes of critical social commentary, explaining that "films can promote revolutionary criticism of social conditions, even of the distribution of property" (231). Like the propaganda and advertisements employed by government and corporations, respectively, visual and written media utilized by the citizenry is *anti*-propaganda that helps to illuminate the path out of the cave of phantasmagoria.

Regarding allegory, it is the process of fragmenting reality *on its surface*—the reality that those in power want the populace to believe—into ruins, so that people can see the reality *within*. This is the destructive power of allegory, as it brings chaos to order as a means of illuminating the truth. In Benjamin's words: "allegory has to do, precisely in its destructive furor, with dispelling the illusion from all 'given order'" (1999, 331; J57,3). In order for allegory to be the most effective, it must not be described, it must be *shown*. Through the use of allegory, the underlying ailments of Japanese society can be brought to the fore and *seen* with one's own eyes, for the "primary interest of allegory is not linguistic but optical" (334; J59,4). In sum, for the citizenry, the allegory is both a political weapon against government power and a "decommodifying" weapon against capitalism.[26] It is a technique used to expose the falsehoods of reality as we know it. James R. Martel explains that according to Benjamin, the allegory, by definition, cannot be misused, for it is an ontological weapon that can only "make truth, not lies."[27]

Advertisements: Bringing the Arcade into the Living Room

As explained in Chapter 2, not unlike 19th-century Paris, 21st-century Tokyo is a haven for the consumer, engulfed in commercialism and practically never-ending consumption. The only major difference between modern-day Tokyo and 19th-century Paris is the degree to which technology has been utilized to enrapture the populace in this "fairy tale" of consumerism. Today, advertisements are abundant and ubiquitous. One will be exposed to advertisements while reading magazines and newspapers, and obviously by watching television. Regarding TV, product placement is no longer limited to commercials in between programs, but has invaded the television show itself. This is especially the case in Japan where an actor in a drama series also is the artist singing the series' theme song.[28] In short, the singer provides the theme song for the TV

---

[26] James R. Martel. "Walter Benjamin and the Use of Allegory" (seminar, San Francisco State University, San Francisco, CA, March 27, 2006).

[27] Ibid.

[28] Examples of this include, but are not limited to: Yuji Oda starring in the drama series "Last Christmas" (2004) and "Jōdan Janai" (2007) and providing the theme songs for each, Aya Ueto starring in "Hitonatsu no Papa e" (2003) and singing the theme for that series, and Kazuya Kamenashi, who was the male lead in "Suppli" (2006), for which the six-member J-Pop group KAT-TUN provided the opening theme; Kamenashi is one of the members of KAT-TUN.

show she or he stars in and the show itself advertises the singer and the song, both indirectly, by the mere fact of featuring both the singer/actor and the song on the show, and directly—at the end of television dramas, it is common for a show to advertise the release date of the single or album featuring the show's theme song (although this is not limited to cases in which the series' actor is the singer). The utilization of the singer/actor's song in the drama series is, in short, a form of "the disguised advertisement" that Benjamin makes note of in Convolute A of *The Arcades Project*, illustrating that ads can be anywhere and everywhere, even where we would least expect them (1999, 54; A9,2).

Of the primary goal of commodity capitalism, Benjamin states: "Giedion shows [...] how the axiom, 'Welcome the crowd and keep it seduced' [...], leads to corrupt architectural practices in the construction of the department store Au Printemps (1881-1889). Function of commodity capital!" (Benjamin 1999, 40; A3,6). In the 21st century, the masses are welcomed and kept seduced even before they enter the store due to the prevalence of ads in their everyday lives. In today's technological and information age, in addition to television and periodicals, ads have infiltrated a medium that has been a fixture of modern life since the 1990s: the Internet. Whether checking email or visiting websites to research a topic for a report, whatever reason a person has for logging on to the 'Net,' she or he will come across at least one advertisement during her or his time online, including ads in the form of the dreaded "pop up" window. The modes of advertisements discussed thus far cover those that bring the ads to one's home or workplace, although living in Tokyo, one cannot escape the onslaught of commercialism when one opens the front door and departs for school or work. Benjamin's description of the ubiquity of print ads (posters) in 19th-century Paris foreshadowed what was to come in future Tokyo: "In the early days of the poster, there was as yet no law to regulate the posting of bills or to provide protection for posters and indeed *from* posters; so one could wake up some morning to find one's window placarded" (65; B2,1). Today, ads are posted inside subway trains (and also televised on trains equipped with monitors) and as one walks down the streets of Shinjuku, Shibuya, or any other of Tokyo's commercial districts (pun intended), one will inevitably be exposed to the onslaught of posters, billboards, and large-screen displays advertising the latest product or utility that a person "must have." We must not forget the largest and most phantasmagoric advertising venue of them all: the store, for it is an advertisement in and of itself. The added bonus of a store being its own mode of advertising is the higher probability of consumers—with the mentality of "I'm already here, I might as well get it"—buying what is being offered for sale.

With advertisements everywhere—in and outside of the home—people living in the modern city are exposed to these phantasmagoric images on a daily basis and as a result are bombarded with messages persuading them to focus their attention on and even *care* about relatively unimportant matters and objects. Thus, instead of concerning oneself with the dilemma of homelessness, the decay of the education system, youth unrest and violence, and continued gender and even ethnic discrimination in Japanese society, one instead places a high priority on purchasing a new Softbank cellular phone, Nintendo Wii, Gatsby hair wax or Shiseido mascara, and keeping abreast of the latest fashion craze or focusing one's attention on the selection of new members for the J-Pop idol group, Morning Musume; all the while ignoring the ongoing debate about Japan's potential military expansion or the movement to revise history, erasing wartime atrocities from the public consciousness. In the end, living in a world of commodity, humanity—mind, body, and perhaps even soul—itself has become a commodity.

The propaganda that the masses are bombarded with every day results in *reaction*. "It thus led to that state of subjection which propaganda, industrial as well as political, relies on," says Benjamin about world expositions and their ability to induce reactionary responses (1999, 18). In Convolute G of *The Arcades Project*, which focuses on advertising, Benjamin notes, "The entertainment industry refines and multiplies the varieties of reactive behavior among the masses. In this way, it makes them ripe for the workings of advertising" (201; G16,7). In short, when all people do is react, they continue to be susceptible to government and corporate propaganda, influenced to think in a certain way, to conform to the viewpoints of society's "overseers." The key to dispersing the phantasmagoria is thus the antithesis to social reaction: namely, *political action*.

The Political Camera: Interaction with Art, Interaction with Politics

Having observed the "sovereign's camera" and its manipulative capabilities in Chapter 1, we shall now look through the viewfinder of the "people's camera" to dispel the phantasmagoria. Benjamin does not perceive of the film as completely evil, as a mode possessed solely by a fascist government. At certain points in "The Work of Art in the Age of Mechanical Reproduction," Benjamin concedes that film may be used in a just fashion. First, as mentioned in the introduction to this chapter, Benjamin says that film can be a means of promulgating social commentary on problems facing society: "We do not deny that in some cases today's films can also promote revolutionary criticism of social conditions, even of the distribution of property" (1988b, 231). Specifically, for Benjamin, film may be utilized by those holding a Marxist-revolutionary standpoint. Even so, he immediately dismisses this thought, asserting, "However, our present study is no more specifically concerned with this than is the film production of Western Europe" (231). Second, Benjamin explains that the audience can, in a sense, engage the film they are watching in a similar manner to how people observe and contemplate works of art such as a painting or sculpture. Benjamin elucidates:

> It is inherent in the technique of the film as well as that of sports that everybody who witnesses its accomplishments is somewhat of an expert. [...] the newsreel offers everyone the opportunity to rise from passer-by to movie extra. In this way any man might even find himself part of a work of art. (231)

On a similar note, the same public engagement with art could be seen in newspapers: "With the increasing extension of the press, which kept placing new political, religious, scientific, professional, and local organs before the readers, an increasing number of readers became writers" (Benjamin 1988b, 232). As readers become writers, they become active as opposed to reactive, and in so doing, they step out of their own private world and into the public world of political life—a world of *action*.

A third positive aspect of film is discussed in Section XIII as Benjamin talks about psychology and human perception. "The film has enriched our field of perception with methods which can be illustrated by those of Freudian theory," says Benjamin (1988b, 235). He then mentions the Freudian slip, stating, "Fifty years ago, a slip of the tongue passed more or less unnoticed" (235). Film, in contrast to a painting or a stage performance, allows us to be more meticulous observers of human behavior due to film's ability to isolate certain behavioral gestures via the "close-ups of the things around us" or "by focusing on hidden details of familiar objects" (236). Whereas a painting represents an object as a *whole*, the film captures images in *fragments*. Prior to the advent of the film:

> Our taverns and our metropolitan streets, our offices and furnished rooms, our railroad stations and our factories appeared to have us locked up hopelessly. Then came the film and burst this prison-world asunder by the dynamite of the tenth of a second, so that now, in the midst of its far-flung ruins and debris, we calmly and adventurously go traveling. (236)

To elaborate, before film was invented, the masses were in a sense trapped within the confines of their own little world, but with the creation and utilization of film, people came out of their self-defined caves and acquired greater awareness of what occurred around them. The concern remains as to whether or not the populace takes advantage of the camera of political power and become cameramen and camerawomen themselves or do nothing and risk becoming servants to the camera of political power. Because of the advances in telecommunication technology, the world is literally at a person's fingertips, available at a push of a button, be it a button of a remote control or computer mouse.

Concerning the use of advertisements by the citizenry, they can utilize this mode of communication to dissipate the mist of the phantasmagoria, to open one another's eyes to the underlying truths of modern society. Benjamin cites the following excerpt written in 1864 of the usage of ads for revolutionary purposes:

Between the February Revolution and the June Revolution: "All the walls were covered with revolutionary posters [...]. There was scarcely a palace or a church on which these notices could not be seen. Never before was such a multitude of placards on view in any city. [...] As the time for the opening of the National Assembly drew near, the language of the posters grew wilder and more passionate. ... The number of public criers increased every day; thousands and thousands of Parisians, who had nothing else to do, became news vendors." (1999, 177; G3,1)

Advertisements thus can be used as "anti-propaganda" for the purpose of increasing public awareness of social issues and problems facing society today and in the future. In other words, advocates for social justice, assuming they acquire the ability to do so on a large enough scale, may co-opt modes of advertising to spread their message across the nation and across the world. Just as ads appear on television, flyers, and billboards promoting designer clothing, electronics, and other "must have" commodity goods, imagine a Japan where the citizenry is constantly exposed to advertisements about violent and suicidal students, modern-day sex slavery, *hentai ojīsan* (roughly translated as "dirty old men") and their teenage prostitutes. Shock advertising is "an age-old gimmick" but still a very effective one (Bennett 2008, 13). According to Steve Hall, founder of AdRants, the imagery presented in these ads is "like a sledgehammer to the face" (Bennett 2008, 13). Jessica Bennett of *Newsweek* explains that such ads, utilizing "deterrence by disgust" do work (13). Again, if not engaged or motivated to take action, at the very least, the public will be made aware of the underlying problems in the supposedly problem-free and orderly society in which they live.

Returning to the subject of the public's interaction with a painting and with a film, Benjamin notes a striking difference. When staring at a motionless painting, viewers are able to excogitate the meaning and relevance of what they are seeing. The same cannot be done when one views a film for the images are constantly moving; it is impossible for viewers to immediately ponder the image before them since they are constantly being bombarded by a slew of new imagery. Benjamin explicates: "Let us compare the screen on which a film unfolds with the canvas of a painting. The painting invites the spectator to contemplation [...] Before the movie frame he cannot do so. No sooner has his eye grasped a scene than it is already changed" (1988b, 238). In other words, film rarely allows for instant contemplation. The quote Benjamin provides from Duhamel articulates this well: A movie is "a pastime for helots, a diversion for uneducated, wretched, worn-out creatures who are consumed by their worries ..., a spectacle that requires no concentration and presupposes no intelligence ..." (239). In short, film permits public involvement and engagement, but rarely a cognitive one. This is indeed a severe problem due to the prevalence of film and television in people's daily lives. As Benjamin states, "A man who concentrates before a work of art is absorbed by it. [...] In contrast, the distracted mass absorbs the work of art" (239). To bring in a salient example, one can take into consideration a news broadcast, which practically consists of one flashing image and one sound bite after the other. People get a 'digest,' an abridged version, of a current event, but before they can process it, a new and unrelated story comes on and grabs their attention. In conclusion, Benjamin declares, "The public is an examiner, but an absent-minded one" (241).

From Charles Baudelaire's *Oeuvres*:

> "The gist of it all, in the eyes of history and of the French people, is that Napoleon III's great claim to renown will have been that he showed how anybody at all, if only he gets hold of the telegraph and the printing presses, can govern a great nation. Anyone who believes that such things can be done without the people's permission is an imbecile." (Benjamin 1999, 315; J48,1)

What Baudelaire says is true. Authoritarian methods of achieving power aside, just because one has control over modes of mass communication does not automatically give one control over the people without their consent. But that is not to deny the fact that in a society with an inattentive or ignorant populace, *anyone*—from the dimwitted to the malevolent—has the potential to acquire and maintain power. "Stupidity is often the ornament of beauty," states Baudelaire (Benjamin 1999, 316; J48,7). The ignorance—the stupidity—of the public is the political leader's "ornament of beauty." As Benjamin says, the public is an absent-minded examiner, and as such, it is susceptible to persuasion and control—but not unless it utilizes film and art technology to combat the phantasmagoria.

Making Progress?

In *The Arcades Project*, Benjamin explores mythic and natural history. The key to understanding these two types of history lies in progress, specifically, the importance placed on progress as it relates to history, and for that matter, as it relates to the future. Susan Buck-Morss presents a work of allegory by John Heartfield entitled "Deutsche Naturgeschichte" (1991, 61). The photomontage of "German Natural History" depicts the heads of German leaders on the bodies of insects. Specifically, the head of Friedrich Ebert, the Weimar Republic's first president, appears on a body of a caterpillar; the head of Paul von Hindenburg, Weimar's last president, on the body of a cocooned caterpillar; and Adolf Hitler's head on the body of a moth. There appears to be a "natural progression" as Ebert metamorphoses into Hitler, as the relatively peaceful Weimar Republic is transformed into the fascist state of the Third Reich. At least, this may be what the German citizenry during the 1930s and early 1940s believed, or were persuaded to believe by the ruling Nazi Party. But in the end, was the Weimar Republic's metamorphosis into the Third Reich really progress? Buck-Morss states that "this progression (on a dying tree branch) is seen as *retrogression,* and 'development' applies only to increasing clarity as to the nature of the beast: the visible mark of the skull, or death's head, in its final Hitlerian form" (60; emphasis added). Overall, in terms of the welfare of a civilization and its populace, if conditions remain the same or get worse—i.e., retrogresses—can a society be said to be making progress? This having been said, is Japan making progress?

Allegory can be used to "de-mythify" what is perceived to be progress and reveal it to be anything but. Allegory employs the grotesque in that it uses grotesque imagery to cast away the phantasmagoria of myth; allegory "disempowers" the intent of the original image.[29] The populace lives in a world of fantasy, but many do not see the horrors that lie hidden within the phantasmagoria. For instance, the Japanese education system is acclaimed as an institution of diligent students. On the surface, this may be true, but if one were to delve deeper, one will discover among the hardworking scholars, a group of suicidal youths and teen prostitutes. When allegory is used to expose this grotesque imagery to the mind's eye, "reality" can never be seen in the same way again. Take for example the allegory of the female body Benjamin presents, showing the grotesque reality of what lies under a woman's sensual exterior:

> "The beauty of the body is merely skin-deep. For if […] men were to see what lies beneath the skin, they would recoil in disgust at the sight of a woman. That well-known charm is nothing but mucus and blood, humors and bile. Just stop to consider what is hidden away in the nostrils, the throat, or the belly: everywhere is filth. And if, in fact, we shrink from touching mucus or dung with even the tip of our finger, how could we ever wish to embrace the sack of excrements itself?" (Benjamin 1999, 402; K7a,4)

The allegory is akin to Pandora's box. When Pandora's box was opened, the evils of humanity were unleashed unto the world. As the box of allegory is opened, the truth is unleashed. That is the power of allegory. Ultimately, when allegorizing "reality" (i.e., what appears to be reality or propagated as reality by society's ruling classes), as Heartfield had done, one may begin to the see the grotesqueness that is the underlying truth behind the façade of myths and lies. While explicating Heartfield's "Deutsche Naturgeschichte," Buck-Morss states:

> Heartfield presents the natural evolution of German political history in the mythical form of a metamorphosis of humans into nature, in order to make the critical point that the belief in evolutionary progress as social history's natural course *is* a myth, in the fully negative sense of illusion, error, ideology. (1991, 62)

An allegory similar to the one employed by Heartfield is applicable to the Japanese case, showing how allegory can be used to illuminate and perhaps

---

[29] James R. Martel. "Walter Benjamin and the Use of Allegory" (seminar, San Francisco State University, San Francisco, CA, March 27, 2006).

even subvert the "Tokyo phantasmagoria." In place of the heads of Ebert, Hindenburg, and Hitler, an allegorist may consider inserting the heads of Prince Naruhiko Higashikuni, Japan's first post-war prime minister; Junichiro Koizumi, Japan's prime minister from 2001 to 2006; and Shinzo Abe onto the bodies of the caterpillar, cocoon, and moth, respectively, to represent the "natural progression"—or retrogression, depending upon one's point of view—of Japan from a shining beacon of the new, of hopes, dreams, and unrealized potential, to a country stagnated—be it because of apathy, hubris, or a combination of the two—by its technological and economic successes; and from a nation that renounced war as a possibility of settling international disputes to one that has made "progress" in expanding its military capabilities.[30]

> "… there is no progress, alas, but merely vulgar revisions and reprints. […] Until now, the past has, for us, meant barbarism, whereas the future has signified progress, science, happiness, illusion! This past, on all our counterpart worlds, has seen the most brilliant civilizations disappear without leaving a trace, and they will continue to disappear without leaving a trace. The future will witness yet again, on billions of worlds, the ignorance, folly, and cruelty of our bygone eras!" Auguste Blanqui, *L'Eternité par les astres: Hypothèse astronomique* (Paris, 1872). (Benjamin 1999, 114; D7; D7a)

According to Blanqui, what is seen as progress is nothing more than an illusion. Barbarisms of the past continue in the present day, in similar forms, but with different names. Despite the advances mankind has made, it is the barbarisms that all humans are capable of that can destroy great civilizations. On its

---

[30] Article 9 of the Japanese Constitution renounces war as an option for resolving international conflicts: "Aspiring sincerely to an international peace based on justice and order, the Japanese people forever renounce war as a sovereign right of the nation and the threat or use of force as means of settling international disputes" (Hayes 2005, 295). Though the constitution, including Article 9, has yet to be amended since its enactment in 1947, in less than three months after Shinzo Abe assumed the Office of Prime Minister, the Diet passed a bill on December 15, 2006, converting the Defense Agency into a Cabinet-level Defense Ministry on January 9, 2007. The upgrading of the agency to a Cabinet-level ministry permits the Defense Ministry to submit proposed legislation to the Diet and to make budget requests directly, although the name change is primarily symbolic in nature. Prime Minister Abe was a strong supporter of the bill elevating the stature of the Defense Agency, which was introduced during his predecessor's tenure as prime minister, and in his first policy speech after succeeding Junichiro Koizumi, Abe promised to make the Defense Ministry a reality. Abe also supported amending the constitution to permit Japan once again to legally have a military in name—and with full offensive capabilities. Hayes explains China's attitude toward Japan's interest in expanding its military capabilities, a view echoed by other Asian nations: "The Chinese view Japan's military activity as greater than that needed for defense, suspecting Japan might be contemplating 'projecting' its military power to other parts of Asia" (232). Ultimately, what is of concern to China and other Asian countries is that if Japan remilitarizes, it will permit the military the option of offensive strikes, and this in turn may open the door to potential expansionist policies by the Japanese government, similar to those precipitating the Second World War.

surface, an outsider looking at Japan may see utopia on earth, but if one were to look deeper—to cut through the phantasmagoria of the Tokyo city lights, the high rate of literacy, and relatively harmonious society of this "beautiful country" [31]—one will discover vagrants scattered among the crowd of affluent Tokyoites, hostile and even suicidal students unable to bear the stresses of bullying or "examination hell," [32] women—Japanese and non-Japanese alike—serving roles similar to those of wartime comfort women, and a government plagued with scandal. Is this progress or merely the illusion of progress? Referencing Baudelaire's poem "Le Crépuscule du matin," Benjamin says, "The morning wind disperses the clouds of myth. Human beings and their affairs are exposed to view. The prerevolutionary dawn glimmers in this poem" (1999, 268; J22,4). Allegory is this morning wind.

The Dialectical Nature of Confucianism: Allegory in Action

Returning to the topic of modes and objects of intention, when allegory is used as a mode of intention, it undoes the original mode of intention applied to the object. Thus, the intent of allegory is to eliminate intention. [33] In the end, we can use allegory to do one of two things: To discover the truth or to discover that there is no truth. Benjamin presents the following example of allegory:

> On the allegorical element. "Dickens ... mentions, among the coffee shops into which he crept in those wretched days, one in St. Martin's Lane, 'of which I only recollect that it stood near the church, and that in the door there was an oval glass plate with COFFEE ROOM painted on it, addressed towards the street. If I ever find myself in a very different kind of coffee room now, but where there is such an inscription on glass, and read it backwards on the wrong side, MOOR EFFOC [...] a shock goes through my blood.' That wild word, 'Moor Eeffoc,' is the motto of all effective realism." (Benjamin 1999, 233; J3,2)

The shock that results from seeing the letters of 'coffee room' in the reverse is a simple illustration of the effect of allegory. One sees strangeness (MOOR EEFFOC) in the ordinary (COFFEE ROOM). It is the shock of encountering the real or simply another interpretation of the real. The letters in 'coffee

---

[31] Shinzo Abe's overall objective after being elected Prime Minister was to make Japan a "beautiful country," stating this goal in a policy speech to the Japanese Diet: "The vision I am aiming for is that of 'a beautiful country, Japan'—a country filled with vitality, opportunity, and compassion, which cherishes a spirit of self-discipline, and is open to the world" (2006).

[32] "Examination hell" refers to the stress and pressure that most senior high school students in Japan experience as they prepare for college entrance examinations. For some, examination hell begins much earlier in life.

[33] James R. Martel. "Walter Benjamin and the Use of Allegory" (seminar, San Francisco State University, San Francisco, CA, March 27, 2006).

room' represent a dialectical image of sorts; on the one hand, it is "normal" when read correctly and on the other hand, it is "grotesque" when the letters are read backwards, but nevertheless, the letters' normal and grotesque forms exist concurrently. The allegory, in short, reveals the existence of this "dark side" of an object or event, and in so doing, illuminates what is underneath the illusory veil of the phantasmagoria.

Earlier I discussed John Heartfield's allegorical "Deutsche Naturgeschichte" and applied a similar image to Japan to allegorize its "natural history." Now I will allegorize the Confucian "virtues" of men and women to present another salient example of how allegory, via its utilization of shocking and/or grotesque imagery, can reveal the hidden reality of various aspects of political and social phenomena in Japan. Like 'coffee room,' Confucianism is dialectical, having one side that all are aware of—the phantasmagoric surface reality—and a side that is concealed, ignored and perhaps even forgotten. Allegory will be used to bring to light this forgotten aspect of Confucianism: the duties and obligations males are expected to show toward women.

Unlike men's "virtues," women's "virtues" are constantly at the forefront in references to Confucianism and the Japanese social hierarchy. Confucian moral values place limits on women's roles in the social structure and *danson johi*, "respect the male, denigrate the female," was a disposition widely held during the Edo period and is still very much intact in modern Japanese society (Yusa 2002, 87). *Onna Daigaku* or *Greater Learning for Women*, a booklet put together by an anonymous author using excerpts from the writings of influential Neo-Confucian scholar Kaibara Ekiken, was widely distributed in the early 1700s (87). The booklet detailed the virtues that females were to abide by throughout their lives. Michiko Yusa gives a brief overview of the content of *Onna Daigaku*, which included the so-called "five ills" and "seven reasons for divorce":

> A girl must practice filial piety toward her father and mother until marriage, and after marriage she must honor her father-in-law and mother-in-law; for a married woman, her husband is her "lord" and she "must serve him with all worship and reverence." Listed as "five ills that afflict women" are "indocility," "discontent," "slander," "jealousy," and "silliness." A woman could be "returned" to her parents' home if found guilty of any of the "seven reasons for divorce," which include disobedience to her father-in-law or mother-in-law, failure to bear children, lewdness, jealousy, and foul diseases such as leprosy; the humiliation of divorce was to be initiated by her husband's family only. (87)

In short, females were expected to obey males throughout the phases in their lives, from adolescence to adulthood. Paul Stapleton summarizes these so-called "virtues" of women: "Daughters should obey their fathers, wives should obey their husbands, and mothers should obey their sons" (1998, 34). But what of men? What are their roles as "good" sons, husbands, and fathers? What are their responsibilities, their virtues?

With regard to the Confucian beliefs and values that have permeated throughout Japanese society, special attention can be paid to obligations within the social hierarchy. Returning to the subject of prostitution and human trafficking, on the basis of Confucian thought, the mindset of Japanese males toward sex workers is generally a demeaning one, in which they see the prostitutes as lower-class humans at best and sub-humans at worst; in the latter case, women become objects, something less than human. Since Japan is a male-dominated society, if ads and educational programs were to be used as allegorical devices to illuminate the problem of human trafficking, they can emphasize men's social obligations toward people of lower rank in the social hierarchy— in this particular case, female sex workers. According to Confucianism, though those of higher social stature are considered superior and should therefore be afforded great respect by their subordinates within and outside of the family, they nevertheless are expected to reciprocate by demonstrating kindness, guidance, and protection to subordinates. Richard D. Lewis states that Confucian social obligations are not just one-sided, "Unequal relationships do not [...] simply imply unlimited advantages for the superiors. While their authority must not be questioned, their obligations are also mandatory" (2006, 487). In short, a superior's duty to a subordinate should be placed at the forefront of allegorical devices, written or visual, against human trafficking. For example, Japanese men who may solicit the service of prostitutes can be made aware of how their actions harm and exploit the women and in so doing behave and act in such a way that runs counter to the Confucian values of helping and protecting those who are hierarchically inferior to them.

Benjamin mentions the dialectics involved in prostitution; specifically, of money and prostitution: "On the dialectical function of money in prostitution. It buys pleasure and, at the same time, becomes the expression of shame" (1999, 492; O1a,4). Prostitution is basically about a man buying pleasure, but at the same time, he also acquires shame, though this latter "purchase" is rarely a subject of discussion in respect to prostitution in Japan. But if it were, it may be a highly effective method of getting men to think about prostitution in a different light since "losing face" is something that is highly frowned upon in Japanese culture (Bramble 2005, 34). Thus, the disgrace of losing face—of being shamed—is a tactic allegorists can employ when targeting men who seek out prostitutes. But in order for this to work efficiently, one must be made aware of his duty toward a subordinate. When a man goes to a prostitute, he has to come to the realization that he is in fact victimizing the woman (or *girl* as is commonly the case with trafficked females) and consequently has failed in his social obligation to protect her. Victor Hugo warns us to beware of the monsters among us: "Cities, like forests, have their dens in which all their vilest and most terrible monsters hide" (Benjamin 1999, 415; L5,4). Tokyo's monsters are not from without but from within, and may very well be its own sons, husbands, and fathers.

"Teddy Bear 2" by Japanese artist Henmaru Machino depicts, using *anime/ manga*-style art, a kneeling girl with stumps in place of arms and legs and a ball gag strapped to her mouth. Behind her is a teddy bear (forcibly?) having anal intercourse with the girl. This work can be considered a commentary on Japan's sexual fetishisms, inclusive of BDSM and *lolicon*, the Japanese abbreviation for "Lolita complex."[34] From a certain perspective, "Teddy Bear 2" is an allegory on child prostitution in Japan. In reference to the quote by Victor Hugo above, a bear is a "monster" that dwells in forests, and the teddy bear is a doll that fathers typically give to their young children; thus, the teddy bear in Machino's art may be interpreted as being representative of a "monstrous" middle-aged man—a father or father figure—having intercourse with a helpless girl—a pre-teen or teen prostitute. This work can also be perceived of as allegorical of the Japanese male's failure to adhere to his Confucian virtues of nurturing and protecting the female, while imposing restrictive "virtues"—as symbolized by the girl's immobile stumps and gagged mouth—on the female. In brief, through an analysis of Machino's "Teddy Bear 2," we allegorize the so-called "virtues" of Confucianism in relation to underage prostitution and are left with grotesque imagery, epitomized by the perversity of a "father" raping a "daughter."

Takashi Murakami: Japan's Baudelaire

Although the politicization of art and the use of allegory have thus far been discussed separately, these two methods of combating the phantasmagoria can be combined as was alluded to in the previous section. One contemporary artist that has done so is Takashi Murakami, whose art is a synthesis of political/ social commentary and allegory. Originally a student of *Nihonga*, a style that blended 19th-century eastern and western art techniques, Murakami developed an interest in the style of art used in *anime* and *manga* due to his fascination with the *otaku* subculture, which has been an influence on his works, which range from painting and sculpting to animation and digital art. Murakami is an artist, and quite a popular one at that, but he sees himself more as a marketer. As Peter Schjeldahl explains in an article on the artist-entrepreneur published in the April 14, 2008 issue of *The New Yorker*, "Warhol as marketer, not as artist, is Murakami's lodestar."

Of "©MURAKAMI," a retrospective of Murakami's art, one notable exhibit, which Schjeldahl (2008) considered to be "the most controversial element in the show," was a Louis Vuitton outlet that sold Murakami-designed merchandise. This fully-functional boutique is the paradigm of Murakami's marketing genius, for its inclusion in the retrospective basically transformed the museum

---

[34] *Lolicon* also refers to a subset of *anime* or *manga* that features girls (or adult females with childlike features) portrayed in an erotic manner.

into a department store with Murakami's artwork serving as advertisements for his products, which were readily available for purchase on site. The "©MU-RAKAMI" retrospective, inclusive of the in-house Louis Vuitton boutique can be seen as an allegory on the commercialization of art, demonstrating that the museum-department store analogy is an ubiquitous occurrence. Places where art and other objects considered to be of intellectual or historical value are housed, museums are not merely venues of display and appreciation of these objects but also venues of *buying* and *selling* of merchandise associated with the objects being exhibited. Most, if not all, museums have souvenir shops (or even online stores) that sell items related to the artwork or artifacts (e.g., lithographs and scaled replicas) on display or items related to the museum itself (e.g., postcards and photobooks).[35] In short, the "©MURAKAMI" retrospective, taken as a whole, is an allegorical representation of how commercialism and consumerism has permeated all facets of modern life, including art.[36]

"Superflat" Japan

Murakami's artistic style is Superflat, a term coined by Murakami himself and having several definitions including, but not limited to: (1) the two-dimensional nature of art featured in *anime* and *manga*, (2) the "flattening out" of the distinction between high and low art, and (3) "the shallow emptiness of [Japan's] consumer culture" (Drohojowska-Philp 2008). Murakami explains that *anime/manga* art piqued his interest because it "was more representative of modern day Japanese life" (Brooklyn Museum 2008a). In turn, Murakami's Superflat art—not only the pieces themselves but how they are *marketed*—is representative of Japanese modern life, referencing and allegorizing aspects of contemporary Japan including the *otaku* subculture, sexual fetishism, and consumerism.

The Louis Vuitton outlet featured in the "©MURAKAMI" retrospective was but one example of allegory employed—purposely or not—by the artist-entrepreneur. Today, Murakami's art, most of which are prepared through collaborative efforts with a team of artists working for his company, Kaikai Kiki Co., Ltd., is marketed to con-

---

[35] As of the writing of this thesis, the Brooklyn Museum (2008b), which is exhibiting the "©MU-RAKAMI" retrospective from April to July 2008, has several items designed by Takashi Murakami available for purchase, including postcards, a mousepad, t-shirts, dolls, and pillows.

[36] Even the title of the exhibition itself, "©MURAKAMI," is an allegory; specifically, it is one that employs appropriation. Midori Matsui, citing Craig Owens's "The Allegorical Impulse: Toward a Theory of Postmodernism," defines appropriation as "an 'allegorical' operation that 'does not invent images but confiscates them,' laying 'claim to the culturally significant, pos[ing] as its interpreter,' adding a new meaning to the original that 'supplants the antecedent one.' In other words, by mimicking an image in contemporary culture, whether part of an advertisement, film, or photography, appropriation makes a small ironic change to reveal a concealed message" (2007, 85). By "confiscating" the copyright symbol © and using it with his name, Murakami, intentionally or unintentionally, allegorizes the artist as seller and art as commodity.

sumers through a variety of mass-produced commodity goods such as handbags and wallets (like those produced by Louis Vuitton), watches, apparel, and toys. Thus, in the manner that the works are presented, they reveal the dialectical nature of works of art: they are simultaneously masterpieces to be admired and contemplated and commodities to be bought and sold. Through his own art and the blatant marketing of it, Murakami allegorizes the activity of the artist, shocking the mindful observer into becoming aware of not only an artwork's dialectical image, but of an artist's as well, i.e., of the artist as seller. In that respect, he can be compared to Baudelaire, an author whose writings allegorized the activity of his contemporaries despite the fact that he himself took part in the phantasmagoria of commodity capitalism. Recall the excerpt from Baudelaire's poem to Sarah mentioned in Chapter 2 in which he writes, "I who sell my thought and would be an author" (Benjamin 1999, 266; J21,2). As an author prostitutes his mind to make a profit via the selling of his written work, an artist prostitutes his talent via the selling of his artwork. Murakami publicly acknowledges this, referring to his own experience and ambitions: "My goal was to make money and build a traditional Japanese house. My parents are from Kyushu and I was raised in Tokyo. My father was a taxi driver and I was poor as a child. I hate the poor life" (Drohojowska-Philp 2008).

So is Takashi Murakami a businessman just trying to make a quick buck (or yen) by any means necessary or is he a social critic? In the end, he may very well be both. *LA Weekly*'s Dani Katz summarizes Murakami's dialectical character: "Branding. Commerce. Money. All these were a part of Murakami's master plan from the get-go, a way to criticize the global culture of consumption while simultaneously basking in it and, yes, profiting from it" (2007, 1). Even though Murakami does benefit from Japan's "ethics of materialism," the imagery in his art can nevertheless be interpreted as critical of Japan's hyperconsumerism and commodity culture. His Superflat movement, to reiterate, seeks to eradicate the division between high and low art, advancing the argument that such a distinction is "nonsensical, artificial, arbitrary" (2). But can't the same be said of materialism and people's self-perceived *need*—a term that I shall distinguish from 'want'—for certain commodity goods over others? Assuming the purpose and function of a particular item are the same regardless of what company's logo is etched on its surface, the distinction between a popular brand-name product and a "no name" product, like that of high and low art, becomes nonsensical, artificial and arbitrary. For instance, strip away the LV monogram from a Louis Vuitton handbag, what do you have? A handbag. Remove the inner lining featuring Burberry's trademarked check pattern from a Burberry trench coat, what's left? A trench coat.

Ms. Hiropon and the Lonesome Cowboy: Allegories of "Ex-timacy"

Consumerism is not the only facet of Japanese society that Murakami allegorizes. Another subject of inquiry in Murakami's art is that of interpersonal relationships (or a lack thereof). Two examples are "My Lonesome Cowboy" and "Hiropon." "My Lonesome Cowboy" is a fiberglass sculpture of a masturbating boy whose discharged semen whirls around to form a lasso above his head and "Hiropon," another fiberglass figure, depicts an excessively large-busted, naked girl holding her nipples which squirt out streams of milk that connect behind her to form a liquid jump rope. These are two separate sculptures, but not unlike Benjamin's writings, they are fragments that can be combined to form a mosaic or montage. "My Lonesome Cowboy" and "Hiropon" both feature sexually vigorous and seemingly happy individuals (they both have smiles on their faces), but they are both alone. These figures offer two self-contained stories that nevertheless intertwine to tell an allegorical tale of loneliness and affectionless sexual activity, two recurring and interrelated themes in the lives of today's Japanese youth.[37]

Because of the fear of rejection or for some other reason, many Japanese males have become *hikikomori*[38] or socially withdrawn, forsaking real flesh-and-blood women for women of the virtual variety such as interactive video-game characters or realistic-looking sex dolls, which are considered by the men who use them to be their "only emotional outlet" (Maeda 2007). Hence, like Murakami's lonesome cowboy "playing" alone, these boys, too, play alone. As for girls, those who feel a lack of fatherly love at home may seek it elsewhere, even from strangers they meet for compensated dates with sex exchanged for cash or merchandise. They may receive compensation, but do they receive love? Without genuine affection from their sex partners, these girls, like their male counterparts and Murakami's "Hiropon" girl, also experience one-sided relationships and "empty love," or what Dick Hebdige would term "ex-timate" (i.e., public and impersonal) relations (2007, 37). In these examples, love and affection are removed from sexual activity, possibly resulting in distorted perceptions of love and the opposite gender. For males interacting with "virtual girls," women become objectified, mere "toys" or "play things."[39] For females, men become nothing more than sources of "pocket money" with the object of one's love disassociated from the living (a potential mate) and linked to the dead (commodity goods). In brief, the sex act itself—be it with an inanimate

---

[37] See p. 38 of ©*MURAKAMI* for a photo of "My Lonesome Cowboy" and "Hiropon" displayed together. Hebdige also inquires into the possible relationship between the two, asking: "are they strangers, siblings, rivals, partners, playmates, co-stars?" (2007, 39)

[38] Hebdige states the number of *hikikomori* in Japan as being approximately one million; this estimate is based on the research of Michael Zielenziger. Japan's NHK TV gives an estimate of between one million and 1.2 million (2007, 42).

[39] Murakami's "Second Mission Project ko²" is a three-piece display depicting a partially naked woman clad in armor who transforms into a futuristic-looking jet fighter. This woman-jet transformer can be seen as symbolic of the objectification of the female body or in general, the transformation of human into machine, into commodity.

sex doll or *enjo kōsai* client—may very well be a loveless one, leaving both the male and female physically and/or psychologically "playing" alone.

Mr. DOB: More Than Just Another "Kawaii" Face

Like any work of art, be it of high or low culture, Murakami's can be critically observed, contemplated and analyzed to find a hidden or even overt meaning or message, as I had done with "My Lonesome Cowboy" and "Hiropon." Another person may look at these two sculptures and offer interpretations distinct from my own. And that is as it should be, for isn't evaluating and finding meaning in art not so dissimilar from the task of the translator who calls into the forest of language and waits for the echo? And as discussed in Chapter 1, the sound of the echo may differ from translator to translator. This variance in sound, or imagery in the case of art, also applies to the "translation" of art's "echoes." Murakami's motifs—smiling flowers, which Laura Fumiko Keehn (2008) conjectures are representative of a "Japanese society [that] has become full of vacant smiles and happy feelings," mushrooms (does Murakami just like drawing mushrooms or do these mushrooms contain a hidden connotation such as being symbolic of the mushroom clouds caused by atomic bombs?), and his Mickey Mouse-like character Mr. DOB[40] (is he just another generic *kawaii* Japanese character or is there a deeper meaning associated with Mr. DOB's appearance and name?)—are all rich with analytical and interpretive potential. Regardless of whatever significance one discovers in Murakami's art, what matters is that one engages in the activity of excogitation. In this regard, Murakami's primary target audience appears to be the youth of Japan.

With his art transformed into relatively affordable stickers, key chains, figurines, and dolls, Murakami is able to appeal to children. Sleeping with a Mr. DOB plushy and sunflower pillow or picking up a *gashapon* (vending machine "capsule toy") on the way to school, all the while wearing a Kaikai & Kiki backpack, the youth of the "Murakami generation" grow up already familiar with these lovable, cutesy characters but as they mature, they may see Mr. DOB and his *kawaii* (cute) compatriots in a completely different light, perhaps as "abstract and fleeting morphing life-form[s] representative of our endless desire to consume" (Yoshitake 2007, 124). As Mika Yoshitake notes in "The Meaning of the Nonsense of Excess," "Murakami's DOB is not a character so much as an erratic manifestation of one's desire to consume and the ceaseless regeneration of this impulse" (126). DOB's dialectical nature, which is representative of Murakami's Superflat art as a whole, is that it is a marketable icon employed

---

[40] Katz comments on the significance of Mr. DOB's name and its relation to Japanese society: "And that folksy little mouse called DOB (Murakami's Disney-esque alter ego and his first and most oft-used character) with his Mickey Mouse gloves, his Sonic the Hedgehog belly, and that silly name—abbreviated Japanese slang for 'Why?'—is really a testament to consumption, an inquiry into market sustainability and an ongoing study in meaninglessness" (2007, 1).

to make Murakami a profit, while also conveying criticism of Japan's material-ism and never-ending consumption.

Humanity's Redemption

Like an elderly woman disguising her physical decay with makeup or the newest fashions, Japan hides its own decay—be it the visible yet invisible va-grants of San'ya, whose very existence is a paradox to the image of Tokyo as the exemplar of the modern, or the erosion of traditional Japanese ethics and morality, replaced by an ethics of materialism and hyper-consumption—under the veil of phantasmagoria. The progress in technology and economic prow-ess that Japan is known for are sheer aesthetics akin to Haussmann's recon-struction of Paris in the late 1800s. Manmade commodities, from fashion to telecommunication technology, are continuously modified and are said to be "improved," though they are, in the end, merely "the unique self-construction of the newest in the medium of what has been" (Benjamin 1999, 64; B1a,2). And as this cycle of the old as the new continues, humanity itself withers and dies, consumed by the very commodity objects they cherish so much. Is there no hope? Can humanity be redeemed? Benjamin believes so. The redemption of humanity lies within the Messiah, who will eviscerate the old "objective" truths of the phantasmagoria and begin the world anew. Benjamin describes this Messianic moment in the conclusion of *The Origin of German Tragic Drama*, a time in which humanity immerses itself in allegory as it once immersed itself in the phantasmagoria, and therein finds redemption:

> In God's world the allegorist awakens. 'Ja / wenn der Höchste wird vom Kirch-Hof erndten ein / So werd ich Todten-Kopff ein Englisch Antlitz seyn.'[41] This solves the riddle of the most fragmented, the most defunct, the most dispersed. Allegory, of course, thereby loses everything that was most peculiar to it: the secret, privileged knowledge, the arbitrary rule in the realm of dead objects, the supposed infinity of a world without hope. All this vanishes with this *one* about-turn, in which the im-mersion of allegory has to clear away the final phantasmagoria of the objective and, left entirely to its own devices, re-discovers itself, not playfully in the earthly world of things, but seriously under the eyes of heaven. (2003, 232)

---

[41] The following translation is offered by John Osborne: "Yea, when the Highest comes to reap the har-vest from the graveyard, then I, a death's head, will be an angel's countenance" (Benjamin 2003, 232).

CONCLUSION

The Messiah Cometh

Of each generation's "Messianic power," Benjamin explains, "The past carries with it a temporal index by which it is referred to redemption. There is a secret agreement between past generations and the present one. [...] Like every generation that preceded us, we have been endowed with a *weak* Messianic power, a power to which the past has a claim" (1988c, 254). This weak Messianic power lies in the past and can only be unlocked by unlocking the past. As those in power look to the past and attempt to reshape it in a way that benefits them, by the same token, the people "have a retroactive force and will constantly call into question every victory, past and present, of the rulers" (255). In the case of Japan, as the government attempts to redefine or even erase the past, the people must make attempts to counter this by keeping the past alive, by exploding the past into the present. This is accomplished by showing that the past and present are essentially one and the same.

Benjamin asseverates that "we [...] would recognize today's life, today's forms, in the life and in the apparently secondary, lost form of that [past] epoch" (1999, 458; N1,11). The present as it relates to the past, says Benjamin, is "purely temporal" but "the relation of what-has-been to the now is dialectical" (463; N3,1). Again, Benjamin plays with modes and objects of intention; one can obviously argue that 'past' and 'present' are merely synonyms for 'what-has-been' and 'now,' respectively, but Benjamin does not see these terms as such. With regard to what-has-been and the now, the two are dialectical in that they can't possibly exist simultaneously—the former resides in the past, while the latter resides in the present—but if we were to remove the temporal association, we will see that what-has-been *is* the now, and vice versa. "In the dialectical image, what has been within a particular epoch is always, simultaneously, 'what has been from time immemorial'" (464; N4,1). As the writings of Walter Benjamin constantly iterate, life is essentially the continuance of the same; the new becomes the old and the old becomes the new. Names may change but forms practically remain unchanged—yesterday's *geisha* is today's hostess; yesterday's comfort woman, today's trafficked sex slave; yesterday's military, today's Self-Defense Force; and yesterday's businessman, today's "corporate soldier."

The Task of the Historian and the Enemy Within

It is the historian—in her or his role as a *translator* of the fragments of the past—that has a great responsibility, for the redemption that people seek in the future is to be found in the past. "Only that historian will have the gift of fanning the spark of hope in the past who is firmly convinced that *even the dead* will not be safe from the enemy if he wins," states Benjamin (1988c, 255). The phrase that "even the dead will not be safe" is a reference to the past—that the past too will not be safe if the "enemy" wins. Of the potential harm that the "enemy" can inflict on the dead, it seems that the Japanese government today has acted in a manner that disturbs the dead. With respect to history, if the conservative members in Japanese government had their way, the dead (e.g., the deceased victims of Japan's expansionist campaigns during World War II) will be forgotten. With so many books already having the history of comfort women removed, it appears, as Benjamin says, "this enemy has not ceased to be victorious" (255). And as a consequence, progress is halted. Benjamin cites the following from Hermann Lotze's *Mikrokosmos*: "Nothing is progress which does not mean an increase of happiness and perfection for those very souls which had suffered in a previous imperfect state" (1999, 478-79; N13,3). The quote, in relation to the Japanese wartime history controversy, affirms that progress cannot be achieved without bringing closure to the past, without giving peace to those, dead or alive, who had suffered through the atrocities of Japan's imperialist "advances" prior to and during the Second World War.

The "enemy" that Benjamin speaks of is a fascist government comprised of those who believe their own propagated path to progress is the correct one and only one that exists, who are unwilling to listen to or tolerate dissent, and who use their ability to define reality—past and present—in such a way as to advance themselves at the cost of the welfare of the people they are expected to serve and defend. This enemy's weapon is commodity capitalism and the euphoria and stresses that it brings to the citizenry. Either infatuated by lust for the acquisition of wealth and commodity goods, or kept occupied by work and making end's meet, the populace are "too busy" with their *social* lives to concern themselves with *political* life. Of the need to work as an inhibitor against dissent and revolutionary tendencies, Benjamin cites: "Today—apart from force of law—you have but one effective guarantee against this revolutionary disposition of the poorer classes: work, the constant necessity of work" (1999, 738; a22a,3). The citizenry's preoccupation with work and its pecuniary rewards thus provide governments with the opportunity, if they so choose, to commit pernicious actions—regardless or how big or small—and have these malfeasances go unaccounted for by the people.

Though the present-day Japanese government is hardly the fascist "enemy" Benjamin refers to in "Theses on the Philosophy of History," there is ever present the *potential* for it to become this "enemy" (although Benjamin would probably not make a distinction between fascism and liberal capitalism). As the adage by Lord Acton states, "Power corrupts and absolute power corrupts absolutely." But this leaves out the fact that the simple *yearning* for power can lead to corruption. Of such desire, Harry Harootunian explains that fascism in Japan during the 1930s "produced an ideology that corresponded to petty bourgeois yearnings [...] that was compatible with imperialism" (2000, xiii). Not only desire—be it for fame, fortune, or power—but the desire to have *more* is essentially an inherent "value" of commodity capitalism.[42] Anyone, even those entrusted with preserving the welfare and security of the state, can have these desires. This yearning for more, combined with the *capability* to achieve more (via technological advances, etc.) can lead governments down the path of fascism. Harootunian notes that in addition to Japan, "during the 1920s and 1930s, virtually every country in the West (including the United States) experienced either a fascist movement or powerful intellectual and cultural impulses that clearly displayed recognizable characteristics associated with fascism" (xii). In the Japanese case, the "petty bourgeois yearnings" Harootunian refers to encompassed "the fetishism of power, strong state, intensive concerns for national culture, corporatism and authoritarianism, technocracy, not to forget militarism" (xiii). These yearnings of late 20th-century Japan are not so different from those of early 21st-century Japan, as can be exemplified by former Prime Minister Shinzo Abe's objectives for making Japan a "beautiful country."

During Abe's relatively short tenure as Prime Minister, which ended when he resigned in September 2007, roughly a year after taking office, the Japanese government passed two laws that ran contrary to Japan's post-war policies. In December 2006, the government approved bills that reinstated patriotic/ nationalist education emphasizing "love of country," "public spirit" and "tradition" into school curriculum, which had been prohibited by the 1947 Fundamental Law of Education; and elevated the Defense Agency to a full-fledged Defense Ministry, defining "overseas missions as a main duty" (Onishi 2006). Onishi (2006) notes, "The laws were passed in keeping with Mr. Abe's drive to instill pride of country among Japanese and claim a larger role for Japan in the world." This desire to claim a "larger role" in the world and other such ambitions, past and present, may thus be referred to with another mode of intention related to commodity culture: greed.

---

[42] Harootunian also says that "capitalism has no really normal state but one of constant expansion" (2000, xv). With constant expansion comes constant desire—constant greed.

Initial support and momentum for the passage of the two laws seem to have been the concoction of political maneuvering to influence public perceptions and opinion. According to Onishi (2006),

> The bills were enacted despite a motion of no confidence by opposition parties against Mr. Abe's administration, two days after a government report revealed that the government had routinely staged town meetings since 2001 to manipulate public opinion. The government had paid people to ask questions and make statements supportive of its policies, including the revision of the education law.

This is but one example of how a government can shape—manipulate—public opinion, and not only that, but exploit members of the public to do so. Therefore, the citizenry should not let the phantasmagoria of "the good life" that modernity has brought blind them to the potential of those in power to yearn for more, to desire *too much*—even if the reasons they give for this yearning is for the "benefit of the people" or creation of a "beautiful nation"—and their ability to manipulate the public, for it is through the citizenry's ignorance or even arrogance (e.g., excessive national pride and hubris) that a democracy can fall and give rise to a fascist state.[43]

Setting the Sails of Revolution

Japan has undoubtedly progressed technologically over the centuries, but there appears to be a lack of *life* progress. Of the average citizen living in a commodity-capitalist society, she or he lives to work and works to live, and so will her or his children when they come of age. Life is an endless cycle, a revolution of "the ever new in the ever same" (Harootunian 2000, x). Life is thus stagnant, but one can go so far as to say that it is even degenerating. Benjamin would argue that the paradise that people in commodity-capitalist societies live in today—be it Japan, the United States, or France—is, for all intents and purposes, hell on earth. On that note, Benjamin may very well have predicted the world of future Tokyo, blinded by the sin of sensation:

> From time immemorial this enigmatic need for sensation has found satisfaction in fashion. […] It is tempting to connect this need for sensation to one of the seven deadly sins, and it is not surprising that a chronicler adds apocalyptic prophecies to this connection and foretells a time when people will have been blinded by the effects of too much electric light and maddened by the tempo of news reporting. (1999, 65-66; B2,1)

This need for sensation that Benjamin describes is the desire for material goods, which is exemplified by fashion and its constant "evolution," appealing to the con-

---

[43] Onishi (2006) alludes to the slippery slope into fascism by stating the concern of those opposed to the passage of the education and Defense Ministry laws: "…to critics, especially of the education law, the moves take Japan further from its postwar pacifist ideals while harking back to vaguely defined, prewar Japanese values."

sumer's need to keep up with or be ahead of the times, and in so doing, conceal her or his own physical (and perhaps even moral) degradation. Are today's *shakaijin* citizens "sound in mind and body, who love truth and justice, esteem individual values, respect labor, have a deep sense of responsibility" (Kempner and Makino 1993, 188) or are they *shinjinrui*, whose value system is deeply entrenched in materialism and the profit motive? One simply has to walk through Tokyo's commercial districts at night to see the many electric lights, whose blinding effects are deeper than most realize. Behind the phantasmagoric curtain of affluence and relatively carefree living, unarguably results of Japan's technological advancement and economic ventures, remain an underclass and an "internal colony," whose very existence is rarely if ever acknowledged in Tokyo's self-made image of itself. Of San'ya, Sister Rita Burdzy, a missionary serving in Japan, says the following: "While in ministry in Sanya, I learned and received much from those who are thought by society to be without value in a country that prizes material possessions and status" (Maryknoll Sisters 2008). Blinded by the image of what Tokyo *should* be, Tokyoites and Japanese in general forget—voluntarily or involuntarily—fellow Japanese who, beneath the surface, are not so different. But it seems that unless a person has a degree from one of the prestigious "Tokyo Six" universities; a position with Toyota or Sony Corporation; or a 45,150 yen Prada wallet or 95,550 yen Gucci handbag; one is excluded from the in-group of what it means to be Japanese, discarded like old goods that have worn out their use value.[44] In the end, what is considered progress is, like fashion, merely superficial and not "skin deep" and has not given Japan a utopia, but only a semblance of utopia.[45] As Benjamin says, "The concept of progress must be grounded in the idea of catastrophe. That things are 'status quo' *is* the catastrophe. [...] hell is not something that awaits us, but this life here and now" (1999, 473; N9a,1). The catastrophe of the status quo is what Benjamin

---

[44] The "Tokyo Six," the six most prominent universities in Tokyo are: Hosei University, Keio University, Meiji University, Rikkyo University, Waseda University, and the University of Tokyo. The 45,150 yen Prada wallet is featured on page 78 of the December 2006 issue of the men's fashion magazine *Gainer*; the 95,550 yen Gucci handbag on page 37 of the January 2007 issue of the women's fashion magazine *Pinky*.

[45] Katz makes a fascinating comparison between the United States and Japan, painting a similar canvas of America's "progress," noting Los Angeles in particular: "Americans too have yet to evolve past a lingering adolescence systematically enforced by an incessant onslaught of distraction, including 9-to-5 slavery, processed food, reality television, celebrity obsession, Web porn and an overwhelming disconnection from nature thanks to religion, technology and 'progress.' [...] And nowhere do we see this stunted growth more than in Los Angeles—the capital of artifice, consumption and distorted values, where pretty people are paid millions to pretend, while half a million of their neighbors go to sleep hungry every night; where wonderful, magical things happen every minute of every day, but go unnoticed in the wash of celebrity gossip and collective complaining; where an exaggerated chasm separates the rich from the poor, the high from the low, and where we nervously await our inevitable 'flattening,' be it by way of the Big One, WWIII, alien invasion or sudden extinction of the cacao bean ..." (2007, 3).

defines as "missed opportunity," while true progress, in his eyes, is "the first revolutionary measure taken" (474; N10,2).

Similar to how Hannah Arendt makes a distinction between social and political life, arguing that people ought to strive for the latter over the former, Benjamin alludes to politics as the means of conquering the endless cycle of the ever-same.[46] "It is the inherent tendency of dialectical experience to dissipate the semblance of eternal sameness, and even of repetition, in history. Authentic political experience is absolutely free of this semblance," states Benjamin (1999, 473; N9,5). In other words, in order to put an end to the seemingly endless cycle—this endless revolution—of no progress, the citizenry must take political action and enact a different kind of perpetual revolution. Benjamin goes on to say, "What matters for the dialectician is to have the wind of world history in his sails. Thinking means for him: setting the sails. What is important is *how* they are set. Words are sails. The way they are set makes them into concepts" (473; N9,6). With the exception of the multi-definitional term of Takashi Murakami's "Superflat," I have primarily discussed the use of different words applied toward one concept (e.g., 'comfort woman' and 'sex slave' as two different "sails" describing the same concept). In concluding this discourse, I shall shift course, reset the sails in another direction so to speak, and use the same word and apply it towards two different concepts. That word is 'revolution'; the two concepts, 'rotation' and 'insurrection.' As the former has already been discussed, we shall now turn our attention to the latter concept.

Having been influenced by Marxist theory, Benjamin supported the revolution of the proletariats, leading to the actual destruction of the old, capitalism, with the new, communism. But as can be seen with China as the exemplar of a "successful" communist state as actualized in the modern world, communism is essentially the perpetuation of the old—capitalism—with its continued abuses against human life. Thus, revolutions, in the traditional sense, i.e., those with the intent of overthrowing by force established political or economic structures, are in effect pyrrhic victories since some form of injustice committed against the masses by those in power continue, regardless of the type of government or economic system. That having been, we should look to another kind of revolution—a revolution of the human mind, and of the human spirit.

There is a Messianic power—a revolutionary spirit—inherent in every human being. The question is whether or not she or he wishes to use it. This Messianic power arises from political action, which is ultimately not about a means to an end, but is an end in itself.[47] It is not necessarily the outcome

[46] See Part II, "The Public and Private Realm," of *The Human Condition* (1998) for Arendt's discussion on the distinction between public (political) and private (social) life.

[47] Gerard P. Heather. "Hannah Arendt: The Human Condition" (lecture, San Francisco State University, San Francisco, CA, April 1, 2005).

of political action that matters, but the *experience* itself. As Gerard P. Heather explains, "Acting morally does not depend on you being effective. Your action is an end in itself. If your criterion for action is 'Will I be effective?' you will probably not do anything."[48] And there are many actions that the average citizen can take. Since the dawn of the age of mechanical reproduction, *anyone* can write and submit an editorial criticizing corporate injustice or government malfeasance or mass produce a work of art drawing attention to pertinent local or global issues. In today's world, with the omnipresence of the Internet and video-sharing websites like YouTube, anyone can become a filmmaker and present social commentary in the form of the motion picture and have her or his work readily available for public viewing—and public engagement.

If the problems of modernity were, instead of being disguised in the illusory image of the "beautiful life," presented for mass consumption in the same ubiquitous manner that commodity goods are, faced with a daily barrage of how "grotesque" reality truly is would likely lead the average person to, at the very least, *think* about society's predicaments, if not persuade the person to political action. A person's attention will thus be drawn away from superficial, material goods and redirected toward relevant *life* issues (e.g., the apathy and delinquency of today's youth, poverty and global warming). At a minimum, more people will begin to concern themselves with taking care of one another, and not about the maintenance of lifeless commodities, to which people have attributed human qualities:

> "With price tag affixed, the commodity comes on the market. Its material quality and individuality are merely an incentive for buying and selling; for the social measure of its value, such quality is of no importance whatsoever. The commodity has become an abstraction. Once escaped from the hand of the producer and divested of its real particularity, it ceases to be a product and to be ruled over by human beings. It has acquired a 'ghostly objectivity' and leads a life of its own. 'A commodity appears, at first sight, to be a trivial and easily understood thing. Our analysis shows that, in reality, it is a vexed and complicated thing, abounding in metaphysical subtleties and theological niceties.' Cut off from the will of man, it aligns itself in a mysterious hierarchy, develops or declines exchangeability, and, in accordance with its own peculiar laws, performs as an actor on a phantom stage. In the language of the commodities exchange, cotton 'soars,' copper 'slumps,' corn 'is active,' coal 'is sluggish,' wheat 'is on the road to recovery,' and petroleum 'displays a healthy trend.' Things have gained autonomy, and they take on human features. ... The commodity has been transformed into an idol that, although the product of human hands, disposes over the human ..." Otto Rühle, *Karl Marx* (Hellerau <1928>). (Benjamin 1999, 181-82; G5,1)

And as commodities "take on human features," people become commodities, defined by *what* they are in terms of their occupation, not by *who* they are, and are therefore dehumanized in the process. As Marx states of the worker:

---

[48] Gerard P. Heather. "Hannah Arendt: The *Vita Activa*" (lecture, San Francisco State University, San Francisco, CA, April 20, 2005).

"... his human qualities exist only insofar as they exist for the capital *alien* to him. ... The worker exists as a worker only when he exists *for himself* as capital; and he exists as capital only when some capital exists *for him*. The existence of capital is *his* existence, ... since it determines the tenor of his life in a manner indifferent to him. ... Production ... produce[s] man as a ... *dehumanized* being." (Benjamin 1999, 652; X1a,1)

The Human Heart

Benjamin says the following about humanity: "Humans as a species reached the end of their development tens of thousands of years ago; but humanity as a species is just at its beginning" (Buck-Morss 1991, 64). The ushering in of this new era of humanity was brought about by technological advancement, which Benjamin discusses in detail in "The Work of Art in the Age of Mechanical Reproduction." The "social *promise* of technology" was the "beginning of a truly human history," a departure from natural history or "ur-history" (64). What is technology's "social promise"? It is the promise of a better society, a society that Benjamin would consider an improvement upon the one in which he lived and the one in which we live today, a society where commodity does not prevail over humanity, where our affection and concern for an inanimate object does not take precedence over our affection and concern for human life—for one another. Can humanity depend on technology to usher in a utopian age? Technology has failed to do so in the past, and has yet to do so in the present.[49] It seems that technology has actually made things worse by "aggravat[ing] exploitation rather than alleviat[ing] the human lot" (Benjamin 1999, 395; K3,5). Baudelaire notes, "I am like a weary man who sees behind him, in the depths of the years, only disillusionment and bitterness, and in front of him only a tempest that brings nothing new" (314; J47a,2).

Instead of technology, Benjamin compels one to look to the revolutionary. Secondly, Benjamin asks one to focus less on progress and more on justice. Technology may bring progress, or at least the semblance of progress, but it is the revolutionary that has the potential to bring about justice in a world of injustice. Just as technology and materialism commodify, it may be the revolu-

---

[49] Most people, laypeople and academics alike, would probably agree that one of the ideals of a utopian society is the abolition of socioeconomic inequality or at the very least, the reduction of the gap between the haves and have-nots. This is impossible to attain in a capitalist society, argues Harootunian: "... capitalism has no really normal state but one of constant expansion; and the expansion requires the permanent production of excess, surplus, in order for it to survive. Part of the price paid for continual expansion is the production of permanent unevenness, permanent imbalance between various sectors of the social formations, the process by which some areas must be sacrificed for the development of others, such as the countryside for the city [...], the colony for the metropole, or even one city for another" (2000, xv). That is the price of capitalism. In the case of Japan, San'ya is the prime example of that price, that sacrifice, required for continual expansion.

tionary spirit—the desire to do good, not for one's self, but for one's posterity—that humanizes. Benjamin explains:

> The activity of a professional revolutionary [...] does not presuppose any faith in progress; it presupposes only the determination to do away with present injustice. [...] Indeed, it is just as worthy of humane ends to rise up out of indignation at prevailing injustice as to seek through revolution to better the existence of future generations. It is just as worthy of the human being; it is also more like the human being. (1999, 339; J61a,3)

In the final analysis, it may not be technology and its possibilities for progress in which we find the promise of a better tomorrow, but perhaps within our own hearts.

> The night revolves in riot 'round the sphinx.
> If we could once lift up its monstrous paw,
> So fascinating to the mind of yesteryear
> (Newton just as much as ancient Hermes),
> Underneath that dark and fatal claw
> We'd find this one word: Love.

> - Victor Hugo, *La Légende des siècles*
> (Benjamin 1999, 301-2; J40a,5)

REFERENCES

Abe, Shinzo. 2006. Policy Speech by Prime Minister Shinzo Abe to the 165th Session of the Diet. Prime Minister of Japan and His Cabinet, September 29. http://www.kantei.go.jp/foreign/abespeech/2006/09/29speech_e.html (accessed March 10, 2008).

Anime News Network. 2007. Tokyo Anime Center to Hold 'Life-Size Figure Party.' November 29. http://www.animenewsnetwork.com/news/2007-11-29/tokyo-anime-center-to-hold-life-size-figure-party (accessed March 5, 2008).

Arendt, Hannah. 1998. *The Human Condition.* 2nd ed. Chicago: University of Chicago Press.

Asahi Shimbun. 2008. INTERVIEW / Kazuo Ogura: People, not events, shaped modern history. April 1. http://www.asahi.com/english/Herald-asahi/IHistory/TKY200804010273.html (accessed April 23, 2008).

Benjamin, Walter. 1988a. "The Task of the Translator." In *Illuminations*, ed. Hannah Arendt and trans. Harry Zohn, 69-82. New York: Schocken Books.

———. 1988b. "The Work of Art in the Age of Mechanical Reproduction." In *Illuminations*, ed. Hannah Arendt and trans. Harry Zohn, 217-52. New York: Schocken Books.

———. 1988c. "Theses on the Philosophy of History." In *Illuminations*, ed. Hannah Arendt and trans. Harry Zohn, 253-64. New York: Schocken Books.

———. 1999. *The Arcades Project.* Ed. Rolf Tiedemann. Trans. Howard Eiland and Kevin McLaughlin. Cambridge: Belknap Press.

———. 2003. *The Origin of German Tragic Drama.* Trans. John Osborne. New York: Verso.

Bennett, Jessica. 2008. "This Is Your Brain on Scary Ads." *Newsweek*, March 3.

Bramble, P. Sean. 2005. *Culture Shock! Japan – A Survival Guide to Customs and Etiquette.* Portland: Graphic Arts Publishing Co.

Brooklyn Museum. 2008a. Exhibitions: ©MURAKAMI. http://www.brooklynmuseum.org/exhibitions/murakami/ (accessed April 28, 2008).

———. 2008b. Shop: ©MURAKAMI. http://shop.brooklynmuseum.org/murakami.html (accessed April 28, 2008).

Brown, J.T. 2003. Enjo Kosai, Anyone? JT Brown on 'compensation dating' in Tokyo. *Modern Japan*, September 19. http://www.hackwriters.com/enjokosai.htm (accessed March 2, 2008).

Buck-Morss, Susan. 1991. *The Dialectics of Seeing: Walter Benjamin and the Arcades Project.* Cambridge: MIT Press.

China Daily. 2007. "Comfort women" historian alarmed. March 12. http://chinadaily.com.cn/world/2007-03/12/content_824829.htm (accessed February 23, 2008).

Chosun Ilbo. 2006. Editorial: Japan Must Teach History as it Happened. September. http://english.chosun.com/w21data/html/news/200609/200609140022.html (accessed February 23, 2008).

———. 2007a. Abe Digs In Heels Over Comfort Women. March 6. http://english.chosun.com/w21data/html/news/200703/200703060026.html (accessed February 23, 2008).

———. 2007b. Abe Remarks on Comfort Women Make Waves. March 5. http://english.chosun.com/w21data/html/news/200703/200703050022.html (accessed February 23, 2008).

———. 2007c. Comfort Women Were 'Raped': U.S. Ambassador to Japan. March 19. http://english.chosun.com/w21data/html/news/200703/200703190023.html (accessed February 23, 2008).

———. 2007d. Editorial: Abe Could Learn a Thing or Two From Murayama. March 22. http://english.chosun.com/w21data/html/news/200703/200703220033.html (accessed February 24, 2008).

———. 2007e. Editorial: Abe Follows in Koizumi's Footsteps. March 5. http://english.chosun.com/w21data/html/news/200703/200703050024.html (accessed February 23, 2008).

———. 2007f. 'No Proof' Army Forced Comfort Women: Abe. March 3. http://english.chosun.com/w21data/html/news/200703/200703030004.html (accessed February 23, 2008).

dialectic. Dictionary.com. *The American Heritage Dictionary of the English Language, Fourth Edition.* Houghton Mifflin Company, 2004. http://dictionary.reference.com/browse/dialectic (accessed March 13, 2008).

Dinan, Kinsey. 2000. *OWED JUSTICE: Thai Women Trafficked into Debt Bondage in Japan.* Human Rights Watch, September 1. Eds. Regan E. Ralph, Joseph Saunders, and Malcolm Smart. http://www.hrw.org/reports/2000/japan/0-ack.htm (accessed March 1, 2008).

Doyon, Paul. 2001. A Review of Higher Education Reform in Modern Japan. *Higher Education* 41: 443-70.

Drohojowsha-Philp, Hunter. 2008. Superflat. *Artnet Magazine*, January 18. http://www.artnet.com/Magazine/features/drohojowska-philp/drohojows-ka-philp1-18-01.asp (accessed April 29, 2008).

Fujiwara, Kaoru. 1998. The world according to the teenagers in Japan: The other side of *Enjo-Kosai*. http://www.tsujiru.net/compass/compass_1998/reg/fuijiwara_ka-oru.htm (accessed March 10, 2008).

Gainer. 2006. Gohoubi honmei wa mainichi tsukaete ibareru mono. December.

Gibson, William. 2003. *Idoru*. New York: Berkley Publishing Group.

Grant, Tiffany. 2004. Tiffany Grant: Current Info. http://www.modusproductions.com/eva-r/tiff/stats.htm (accessed February 18, 2008).

Harootunian, Harry. 2000. *Overcome by Modernity: History, Culture, and Community in Interwar Japan*. Princeton: Princeton University Press.

Harvey, Ty. 2004. Will the slum remain the same? A visit to San'ya, home to Japan's underclass, where community values are on the auction block. http://findarticles.com/p/articles/mi_m0NTN/is_54/ai_115408929 (accessed February 28, 2008).

Hayes, Louis D. 2005. *Introduction to Japanese Politics*. 4th ed. Armonk: M.E. Sharpe.

Heather, Gerard P. "Hannah Arendt: The Human Condition." Lecture, San Francisco State University, San Francisco, CA, April 1, 2005.

———. "Hannah Arendt: The *Vita Activa*." Lecture, San Francisco State University, San Francisco, CA, April 20, 2005.

———. "Political Theory as Action." Lecture, San Francisco State University, San Francisco, CA, April 15, 2005.

Hebdige, Dick. 2007. "Flat Boy vs. Skinny Boy." In ©*MURAKAMI*, ed. Paul Schimmel, 14-51. New York: Rizzoli International.

Human Rights Watch. 1994. *A MODERN FORM OF SLAVERY: Trafficking of Burmese Women and Girls into Brothels in Thailand*. January 31. http://www.hrw.org/reports/1993/thailand/ (accessed March 1, 2008).

Japan-guide.com. 2000. Valentine's Day and White Day. http://www.japan-guide.com/topic/0003.html (accessed March 11, 2008).

Jong, Son-U. 2007. Editorial: The Twisted Logic of Mr. and Mrs. Nakaya-ma. *Chosun Ilbo*, March 6. http://english.chosun.com/w21data/html/news/200703/200703060010.html (accessed February 23, 2008).

Katz, Dani. 2007. RESISTANCE IS FUTILE: How Takashi Murakami remade the world in his image. *LA Weekly*, October 24. http://www.laweekly.com/art+books/art/resistance-is-futile/17536/LA/?page=1 (accessed April 29, 2008).

Keehn, Laura Fumiko. 2008. Takashi Murakami. *Swindle*. http://swindlemagazine.com/issueicons2/takashi-murakami/ (accessed April 29, 2008).

Kempner, Ken and Misao Makino. 1993. Cultural Influences on the Construction of Knowledge in Japanese Higher Education. *Comparative Education* 29.2: 185-99.

Kingston, Jeff. 2005. *Japan's Quiet Transformation: Social change and civil society in the twenty-first century*. New York: RoutledgeCurzon.

Kioka, Yasumasa. 2003. Dating Sites and the Japanese Experience. National Police Agency of Japan, March. http://www.iajapan.org/hotline/mobilepdf/4_KIOKA.pdf (accessed March 2, 2008).

Kono, Yohei. 1993. Statement by the Chief Cabinet Secretary Yohei Kono on the result of the study on the issue of "comfort women." Ministry of Foreign Affairs of Japan, August 4. http://www.mofa.go.jp/policy/women/fund/state9308.html (accessed February 22, 2008).

Lewis, Richard D. 2006. *When Cultures Collide: Leading Across Cultures*. Boston: Nicholas Brealey Publishing.

Maeda, Toshi. 2007. Japan's lonely hearts turn to dolls for sex, company. *Reuters*, July 18. http://www.reuters.com/article/newsOne/idUSSP10422420070718?sp=true (accessed April 29, 2008).

Martel, James R. "Susan Buck-Morss's *The Dialectics of Seeing*." Seminar, San Francisco State University, San Francisco, CA, April 10, 2006.

———. "Walter Benjamin and the Use of Allegory." Seminar, San Francisco State University, San Francisco, CA, March 27, 2006.

———. "Walter Benjamin's 'Theses on the Philosophy of History.'" Seminar, San Francisco State University, San Francisco, CA, February 20, 2006.

Maryknoll Sisters. 2008. Missioners: Sister Rita Burdzy. http://www.mklsisters.org/index.php?option=com_content&task=view&id=197&Itemid=1 (accessed March 28, 2008).

Matsui, Midori. 2007. "Murakami Matrix: Takashi Murakami's Instrumentalization of Japanese Postmodern Culture." In ©*MURAKAMI*, ed. Paul Schimmel, 80-127. New York: Rizzoli International.

Matsumoto, David. 2002. *The New Japan: Debunking Seven Cultural Stereotypes*. Boston: Intercultural Press.

McDougall, Gay J. 1998. An Analysis of the Legal Liability of the Government of Japan for "Comfort Women Stations" Established During the Second World War. United Nations Commission on Human Rights, June 22. http://www.unhchr.ch/ Huridocda/Huridoca.nsf/7fba5363523b20cdc12565a800312a4b/3d25270b5fa3ea 998025665f0032f220?OpenDocument#Appendix (accessed February 23, 2008).

Ministry of Internal Affairs and Communications of Japan. 2003. Statistics Bureau. The 2002 Employment Status Survey: Summary of Results. http://www.stat.go.jp/ english/data/shugyou/2002/kakuhou/youyaku.htm (accessed March 14, 2008).

―――. 2007. Statistics Bureau. Table 16-29: Starting Salary of New School Graduates by Industry, Size of Enterprise and Academic Career. *Japan Statistical Yearbook 2008*. http://www.stat.go.jp/data/nenkan/zuhyou/y1629000.xls (accessed March 14, 2008).

Nakamura, Akemi. 2007. COMFORT WOMEN: Were they teen-rape slaves or paid pros? *Japan Times*, March 20. http://search.japantimes.co.jp/cgibin/nn2007 0320i1.html (accessed February 21, 2008).

New York Times. 2007. Editorial: No Comfort. March 6. http://www.nytimes. com/2007/03/06/opinion/06tues3.html (accessed February 23, 2008).

Nimura, Janice P. 2004. A Review of 'The Tale of Genji' translated by Royall Tyler. June 8. http://dspace.anu.edu.au/bitstream/1885/42042/1/tyler_review.html (accessed February 18, 2008).

Onishi, Norimitsu. 2006. Japanese Lawmakers Pass Two Laws That Shift the Nation Away From Its Postwar Pacifism. *New York Times*, December 16. http://www. nytimes.com/2006/12/16/world/asia/16japan.html (accessed March 31, 2008).

―――. 2007a. Historian documents Japan's role in sex slavery. *International Herald Tribune*, March 30. http://www.iht.com/articles/2007/03/30/asia/japan.php (accessed February 20, 2008).

―――. 2007b. Japan's Textbooks Reflect Revised History. *New York Times*, April 1. http://www.nytimes.com/2007/04/01/world/asia/01japan.html (accessed February 19, 2008).

Pinky. 2007. GUCCI no shinsaku komono de karei naru LADY rukku! January 1.

Polaris Project. 2004. Who are the Traffickers? *Human Trafficking 101*. http://www. humantrafficking.com/humantrafficking/trafficking_ht3/who_traffickers.htm (accessed March 1, 2008).

prostitute. Dictionary.com. *The American Heritage Dictionary of the English Language, Fourth Edition*. Houghton Mifflin Company, 2004. http://dictionary.reference. com/browse/prostitute (accessed March 7, 2008).

Ritzer, George. 2005. *Enchanting a Disenchanted World*. 2nd ed. Thousand Oaks: Pine Forge Press.

Rouse, W.H.D., trans. 1983. *The Republic*. In *Great Dialogues of Plato*, eds. Eric H. Warmington and Philip G. Rouse, 118-422. New York: Penguin Books.

Rousseau, Jean Jacques. 2006. *The Confessions of Jean Jacques Rousseau*. Ed. David Widger. Project Gutenberg Literary Archive Foundation. http://www.gutenberg.org/ files/3913/3913-h/3913-h.htm (accessed March 16, 2008).

Schjeldahl, Peter. 2008. BUYING IT: A Takashi Murakami retrospective. *New Yorker*, April 14. http://www.newyorker.com/arts/critics/ artworld/2008/04/14/080414craw_artworld_schjeldahl?currentPage=all (accessed April 24, 2008).

Stapleton, Paul. 1998. *Beneath the Surface*. Ed. Akira Ito. Tokyo: Seibido.

Tabuchi, Hiroko. 2007. Historians Find New Proof on Sex Slaves. *Washington Post*, April 17. http://www.washingtonpost.com/wp-dyn/content/article/2007/04/17/AR2007041701380.html (accessed February 20, 2008).

Yajima, Midori. 1992. "Japanese Women." In *Cultural Tradition in Japan Today*, 62-73. Tokyo: Maruzen Co. Ltd.

Yamada, Alice. 1996. Teenage Prostitution in Japan. *Trincoll Journal*. http://www.trincoll.edu/zines/tj/tj02.06.97/articles/inter.html (accessed March 10, 2008).

Yamada, Masahiro. 1998. *The Japanese Family in Transition*. Tokyo: Foreign Press Center.

Yoshida, Reiji. 2007. Sex slave history erased from texts; '93 apology next? *Japan Times*, March 11. http://search.japantimes.co.jp/cgi-bin/nn20070311f1.html (accessed February 23, 2008).

Yoshitake, Mika. 2007. "The Meaning of the Nonsense of Excess." In ©*MURAKAMI*, ed. Paul Schimmel, 111-27. New York: Rizzoli International.

Yusa, Michiko. 2002. *Japanese Religions*. Ed. Ninian Smart. London: Routledge.